Springer Tracts in Modern Physics 102

Editor: G. Höhler
Associate Editor: E. A. Niekisch

Editorial Board: S. Flügge H. Haken J. Hamilton
H. Lehmann W. Paul

Springer Tracts in Modern Physics

---

* denotes a volume which contains a Classified Index starting from Volume 36.

Gustav Kramer

Theory of Jets in Electron-Positron Annihilation

With 86 Figures

Springer-Verlag Berlin Heidelberg GmbH

Professor Dr. Gustav Kramer

II. Institut für Theoretische Physik der Universität Hamburg, Notkestrasse 85
D-2000 Hamburg 52, Fed. Rep. of Germany

Manuscripts for publication should be addressed to:

Gerhard Höhler

Institut für Theoretische Kernphysik der Universität Karlsruhe
Postfach 6380, D-7500 Karlsruhe 1, Fed. Rep. of Germany

*Proofs and all correspondence concerning papers in the process of publication
should be addressed to:*

Ernst A. Niekisch

Haubourdinstrasse 6, D-5170 Jülich 1, Fed. Rep. of Germany

ISBN 978-3-662-15730-5   ISBN 978-3-540-38821-0 (eBook)
DOI 10.1007/978-3-540-38821-0

Library of Congress Cataloging in Publication Data. Kramer, Gustav, 1932–. Theory of jets in electron-positron annihilation. (Springer tracts in modern physics; 102) Bibliography: p. Includes index. 1. Electron-positron interactions. 2. Jets (Nuclear physics). 3. Quantum chromodynamics. 4. Perturbation (Quantum dynamics). I. Title. II. Series. QC1.S797 vol. 102   530s [539.7'54]   83-27095 [QC793.5.E628]

Preface

In the last ten years our understanding of the nature of the strong interaction force has changed considerably. To a large extent this has its origin in our belief that with quantum chromodynamics (QCD), the gauge field theory of quarks and gluons, we have now available a solid, well defined, theory. In the sixties the quark model played already a great role for classifying hadron states but it lacked this theoretical foundation which, we think, now exists with QCD. If we try to remember, which experimental discoveries smoothed the way for QCD, presumably we would consider: (i) the discovery of the scaling behaviour of deep inelastic lepton-nucleon scattering and its interpretation through the parton model, (ii) the discovery of additional quark flavours, the charm and the bottom quark, and the interpretation of corresponding hadron states in terms of simple potential models based on the confinement hypothesis and (iii) the discovery of quark and gluon jets in electron-positron annihilation into hadrons. On the theoretical side it was important to recognize that non-abelian gauge theories are renormalizable and that quantum chromodynamics is asymptotically free. These two properties opened the road for the application of perturbative theory which was so successful in quantum electrodynamics. Furthermore the originally apparent contradiction, the quark-gluon interaction being very strong, i.e. so strong that quarks and gluons are always confined, on one side, and the appearance of almost free quarks and gluons inside hadrons, as revealed in deep inelastic scattering experiments, on the other side, could be resolved.

In the meantime we have learned how to apply perturbative QCD to various reactions at high energies: e^+e^- annihilation, two-photon processes, deep inelastic lepton-nucleon scattering, $\mu^+\mu^-$ production in hadron-hadron collisions (Drell-Yan process) and production of particles with high transverse momenta and of jets in hadron-hadron collisions.

We can distinguish two fields for applying QCD perturbation theory which are only indirectly connected. One is predicting the Q^2 evolution of structure functions of hadrons (or decay functions of quarks and gluons) in deep-inelastic lepton-nucleon scattering and in hadron-hadron collisions. This Q^2 evolution essentially follows from summed perturbation theory in the leading logarithm approximation. The other field consists of the theoretical analysis of jet phenomena in terms of QCD perturbation theory at fixed order. For the first subject, the theory of structure and decay functions many reviews exist. We mention a few: /Altarelli, 1982; Buras, 1980; Dokshitser, D'yakonov, and Troyan, 1980; Ellis and Sachrajda, 1979; Field, 1978; Marciano and Pagels, 1978; Petermann, 1979; Pennington, 1983; Politzer, 1974; Reya, 1981; Ross, 1981; Sachrajda, 1982; Söding and Wolf, 1981/. In some of these reviews also jet phenomena are considered.

It seems generally agreed upon that e^+e^- annihilation into hadrons is the best laboratory to investigate the production of hadron jets. In e^+e^- annihilation we have a well defined, hadron free, initial state which allows us to study the final state undisturbed from effects of initial state hadrons. In deep inelastic lepton-nucleon scattering and even more so in hadron-hadron processes we always have hadrons in the initial state (with more or less known structure functions) producing beam and target jets which overlap with the perturbative QCD jets we are interested in. So far most of the empirical information on jets comes from e^+e^- annihilation experiments. But also on the theoretical side the analysis of jets in e^+e^- annihilation appears to be very much advanced. Here, many results on higher orders in QCD perturbation theory have been obtained in the last three years. Therefore, in this review we shall restrict ourselves to a representation of the theoretical considerations for the analysis of hadron jets produced in e^+e^- annihilation. We hope that the procedures outlined for this particular process may be useful also for the interpretation of jet phenomena in the other more complicated reactions mentioned above. Some earlier reviews of QCD jets are /Hoyer, 1980; Kramer, 1980; Schierholz, 1979,1981; Walsh, 1980/.

Hamburg, Februar 1984   *G. Kramer*

Contents

1. Introduction

1.1 Quarks with Colour

Since Gell-Mann /1964/ and Zweig /1964/ introduced quarks as the elementary build-
ing blocks of all hadrons our understanding of the complex hadronic world of protons, neutrons, pions, kaons and all the other strongly interacting particles has
increased remarkably. Quarks are spin 1/2 particles, so that a quark q and an antiquark $\bar{q}$ build mesons ($q\bar{q}$) and three quarks make baryons (qqq).

To explain the whole hadron spectrum as it exists today we need five quarks of
different flavour: u, d, s, c and b quarks. They differ in their masses and in
their properties concerning electromagnetic and weak interactions. Their quantum
numbers I, I_3, s, c, b, Q and B are listed in Table 1.1. The u, d quarks transform
as a doublet under an almost exact SU(2) flavour group, the u, d, s transform as a
triplet under an approximate SU(3) flavour group, u, d, s, c as a quartet under a
broken SU(4) flavour group and so on. The quark flavours, charges and baryon numbers determine the flavour of all hadrons, their isospin quantum numbers I, I_3,
their strangeness s, their charm c, their bottomness b, their charge Q and their
baryon number B.

The masses of u, d quarks are approximately 10 MeV, the mass of the s quark is
near 150 MeV, of the c quark 1200 MeV and of the b quark 5000 MeV. These masses are
not very well known. They are just parameters which quarks would have as masses if
they could be produced as free particles. Since this is not the case, the masses
cannot be measured directly and their values depend somewhat on the more indirect
definition of the mass parameters.

In addition every quark u, d, s, c and b appears in three distinct states, a red,
a green and a blue quark — a property we call colour /Greenberg, 1964; Han and Nambu, 1965; Gell-Mann, 1972/. That quarks must have another degree of freedom in addition to spin and flavour was revealed by the symmetry problem of baryon ground states
made out of quarks. For example, the lowest mass, spin 3/2, states of three apparently identical quarks, three u's, d's or s quarks depending whether one considers the
Δ^{++}, Δ^- or Ω^- baryon, can be totally symmetric in their spin, spatial and flavour

Table 1.1. The five quarks and their flavour quantum numbers: isospin, strangeness, charm, bottomness, charge and baryon number

Quarks	I	I_3	s	c	b	Q	B
u	1/2	1/2	0	0	0	2/3	1/3
d	1/2	-1/2	0	0	0	-1/3	1/3
s	0	0	-1	0	0	-1/3	1/3
c	0	0	0	1	0	2/3	1/3
b	0	0	0	0	-1	-1/3	1/3

properties, as one expects for the ground state and yet satisfy the Pauli principle, i.e. obey Fermi-Dirac statistics. The antisymmetry of the wave function comes from the colour wave function which is the antisymmetric combination of a red, green and blue quark of the particular flavour u, d, or s. A meson is a linear superposition of red-antired, green-antigreen and blue-antiblue states. So, hadrons, being observable states, are colour singlets although each is built of coloured quarks. This construction explains why mesons, being colour singlets, behave as if made from just one $q\bar{q}$ pair and baryons as if made from a qqq configuration. Further hints for the colour of quarks come from the observed decay rate for $\pi^0 \to \gamma\gamma$ and the cross section for e^+e^- annihilation into hadrons which will be discussed in detail in Chap.2. These physical observables count the number N_c of quarks of each flavour and the experimental data tell us that this number N_c is equal to three.

That the idea of quarks is more than a tool for constructing hadrons was made apparent by the experiments on deep inelastic lepton scattering which started at the Stanford Linear Accelerator. In these experiments, a high momentum probe, a virtual photon or weak virtual quanta $W^\pm$ or Z, hits a nucleon (Fig.1.1). If its momentum is high enough its wavelength is smaller than the size of the nucleon and we expect it to probe the constituents of the nucleon. This is what the scattering experiments really have shown. The scattering of high energy leptons occurs in such a way as if

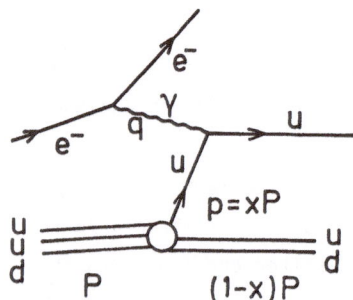

Fig.1.1. Parton model diagram for inelastic electron-proton scattering $e^- + P \to e^- + u$-quark + (ud)-diquark

there are constituents — partons — inside the hadron which are freely moving, point-like objects /Bjorken, 1967; Feynman, 1969; Bjorken and Paschos, 1969/. These partons are found to have spin 1/2 and all the other properties of quarks. Therefore the lepton scattering can be described such that the virtual photon, $W^{\pm}$ or Z with squared $q^2 < 0$ scatter on the quasi-free quarks bound in the nucleon. The final state consists then of two jets, the current jet which is the quark (q) jet and the target jet equal to the diquark (qq) jet. But, however hard the quarks inside the nucleon are hit, the quarks never appear asymptotically as free particles. The quark and the diquark jet must fragment into hadrons which then are observed in the detector.

The fact, that quarks never seem to come out as free particles, whereas hadrons do, is in accord with the colour assignments discussed above. Only colour neutrals, i.e. hadrons, are asymptotic states, All colour non-singlets, i.e. quarks, diquarks etc., cannot appear asymptotically. They always must be bound into hadrons, i.e. they are confined. This means that quarks are strongly bound inside hadrons. What is responsible for this very strong binding? Evidence that a nucleon contains not just three quarks came from the experimental fact that in deep inelastic lepton scattering the charged constituents of a nucleon carry only half of its momentum. An electrically neutral parton carries the rest. This is identified with the gluon. Quarks are assumed bound by exchanging gluons. For example a red quark interacts with a green one by exchanging a red-antigreen gluon. These gluons are flavour neutral and do not participate in the electro-weak interactions.

Since quarks have three colours, there are nine types of gluon. All except one — the singlet gluon — mix under colour transformations. The singlet gluon may have a coupling to quarks of strength independent of the other eight. This is set to zero. The remaining eight gluons transform as the adjoint representation of colour SU(3). Gluons are assumed to have spin one. This has the effect that the force between $q\bar{q}$ is attractive, as it is needed for binding in a meson, but repulsive between qq. This leads us directly to quantum chromodynamics, the gauge theory of quarks and gluons.

1.2 The Lagrangian of Quantum Chromodynamics

Quantum chromodynamics (QCD) /Fritzsch and Gell-Mann, 1972/ is the theory which describes the interaction of a triplet of coloured quarks with an octet of vector gluons by a Yang-Mills gauge theory /Yang and Mills, 1954/. The quark fields are spinors $q_c(x)$ which transforming as the fundamental representation of SU(3) have colour quantum numbers c = 1, 2, 3. The gluon fields $A_\mu^a(x)$ transforming according to the adjoint representation have a = 1, 2, ..., 8. The SU(3) colour transformations are generated by 3×3 matrices T^a (a = 1, 2, ..., 8) ($T^a = \lambda_a/2$, where the

λ_a's are the well known Gell-Mann matrices /Gell-Mann, 1962/). They obey the commutator relations

$$[T^a,T^b] = if^{abc}T^c \qquad (1.2.1)$$

with f^{abc} being the structure constants of SU(3). The Lagrangian density for QCD has the following form

$$\mathcal{L} = -\frac{1}{4}F^a_{\mu\nu}F^{\mu\nu,a} + \bar{q}(i\gamma_\mu D^\mu - m)q \qquad (1.2.2)$$

where the field strength tensor

$$F^a_{\mu\nu} = \partial_\mu A^a_\nu - \partial_\nu A^a_\mu + gf^{abc}A^b_\mu A^c_\nu \qquad (1.2.3)$$

and the covariant derivative

$$D_\mu = \partial_\mu - igT^a A^a_\mu(x) \quad . \qquad (1.2.4)$$

g is the bare coupling constant of the theory and m the bare mass of the quark field q(x). The gluons are massless. By splitting (1.2.2) into a non-interacting part and an interacting part one can read off the Feynman rules from which one can calculate quark-gluon processes perturbatively in g.

The Lagrangian (1.2.2) is invariant under the infinitesimal local gauge transformation defined by $\Theta^a(x)$

$$q(x) \rightarrow q(x) + iT^a\Theta^a(x)q(x)$$
$$\bar{q}(x) \rightarrow \bar{q}(x) - i\bar{q}(x)T^a\Theta^a(x) \qquad (1.2.5)$$
$$A^a_\mu(x) \rightarrow A^a_\mu(x) - f^{abc}\Theta^b(x)A^c_\mu(x) + \frac{1}{g}\partial_\mu\Theta^a(x) \quad .$$

The requirement of local gauge invariance leads to the unique Lagrangian (1.2.2) which severely restricts the otherwise possible interaction terms between quarks and gluons. This gauge invariance is also crucial to make the theory renormalizable /t'Hooft, 1971/ and so yields sensible predictions for physical processes at high energies.

Locally gauge invariant theories like QCD are difficult to quantize because the fields $A^a_\mu(x)$ are gauge quantities and therefore exhibit extra non-physical degrees of freedom which must be dealt with. The most convenient procedure for quantization is Feynman's path integral formalism. For review of this topic see Abers and Lee

/1973/, Zinn-Justin /1975/, Becher, Böhm and Joos /1981/ and Itzykson and Zuber /1980/.

The structure of QCD is similar to that of Quantum Electrodynamics (QED), the only successful field theory we have. In QED, which is an abelian gauge theory, the right hand side of (1.2.1) vanishes and the charged matter fields transform under gauge transformations by simple phase transformations, i.e. U(1) transformations. In QCD, being a non-abelian generalization, the quarks transform under the more complicated SU(3) colour group and the vector bosons, the gluons, now carry colour charge too.

The Lagrangian (1.2.2) is written in terms of the so-called unrenormalized fields $q(x)$ and $A_\mu^a(x)$. The calculation of scattering matrix elements or other physical quantities yield finite results only if the theory is renormalized. This means, the infinities of the theory are absorbed into the basic constants of the theory such as coupling constants and masses which are renormalized to their finite physical values. Therefore these coupling constants and masses must be given and cannot be calculated in this theory. The technique for renormalizing perturbative QCD is well known from QED. The fields are multiplicatively renormalized, i.e. one defines renormalized fields q_r and $A_{\mu,r}^a$

$$q = Z_2^{1/2} q_r$$
$$A_\mu^a = Z_3^{1/2} A_{\mu,r}^a \quad .$$

(1.2.6)

Z_2 and Z_3 are renormalization constants of the quark and the gluon field respectively. In terms of the renormalized fields the QCD Lagrangian has the following form

$$\mathcal{L} = -\frac{1}{4} Z_3 (\partial_\mu A_\nu^a - \partial_\nu A_\mu^a)_r^2 - \frac{\kappa}{2} Z_3 (\partial^\mu A_\mu^a)_r^2$$

$$+ \tilde{Z}_3 (\eta^{a^+} \partial^2 \eta^a)_r - Z_3^{3/2} g f^{abc} (A_\mu^a A_\nu^b \partial^\mu A^{\nu,c})_r$$

$$-\frac{1}{4} Z_3^2 g^2 f^{abc} f^{ab'c'} (A_\mu^b A_\nu^c A^{\mu,b'} A^{\nu,c'})_r + \tilde{Z}_3 Z_3^{1/2} g f^{abc} [\eta^{a^+} \partial^\mu (A_\mu^b \eta^c)]_r$$

$$+ Z_2 (\bar{q}(i\gamma^\mu \partial_\mu - m)q)_r + g Z_2 Z_3^{1/2} (\bar{q} T^a \gamma^\mu A_\mu^a q)_r \quad .$$

(1.2.7)

This Lagrangian is complete. It contains also the gauge fixing term proportional to κ, familiar from QED, which is required to insure a proper quantization procedure. $\kappa = 0$ being the Landau and $\kappa = 1$ is the Feynman gauge. The first two terms in (1.2.7) determine the gluon propagator. The gluon propagator, however, contains too many degrees of freedom for a physical massless vector particle. So it includes an unphysi-

5

cal scalar component which must be removed. Removal of these unphysical states is achieved by adding a Fadeev-Popov ghost term involving $\eta^a(x)$. These occur at all places where there are gluon loops. The propagator of the ghost field is also read off from (1.2.7) as well as the coupling of the ghost field with the gluon field. The ghost field has the renormalization constant Z_3.

In momentum space the free propagators have the following form

quarks propagator $\qquad a \xrightarrow{\hspace{2cm}} b \qquad \dfrac{i}{(2\pi)^4}\delta_{ab}\dfrac{1}{\gamma p - m}$

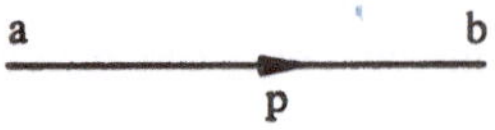

gluon propagator $\qquad \mu,a \,\text{〰〰〰〰}\, \nu,b \qquad \dfrac{i}{(2\pi)^4}\delta_{ab}\dfrac{1}{k^2}\left(-g_{\mu\nu} + \dfrac{k_\mu k_\nu}{k^2} - \kappa\,\dfrac{k_\mu k_\nu}{k^2}\right)$

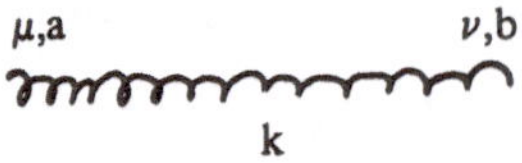

ghost propagator $\qquad a \,\text{- - - - - - -}\, b \qquad \dfrac{i}{(2\pi)^4}\delta_{ab}\dfrac{1}{k^2} \qquad$ (1.2.8)

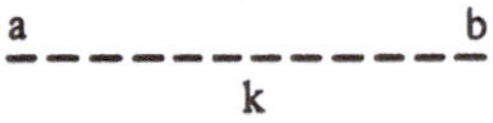

The gluon and the ghost are massless. Of course the ghost field does not appear as an external particle. The vertices, i.e. the quark-quark-vertex, the three-gluon vertex, the four-gluon vertex and the ghost-ghost-gluon vertex are also obtained from (1.2.7). They are represented in momentum space in Fig.1.2 together with the

$= ie\,\gamma_\mu (2\pi)^4 \delta(k + p_2 - p_1)$

(a)

$= ig\,\gamma_\mu\, T_a (2\pi)^4 \delta(k + p_2 - p_1)$

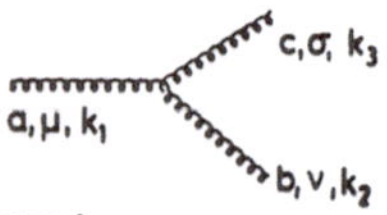

$= g\, f^{abc}\, (g_{\mu\nu}\,(k_1 - k_2)_\sigma + \text{cyclic}) \cdot (2\pi)^4\,\delta(k_1 + k_2 + k_3)$

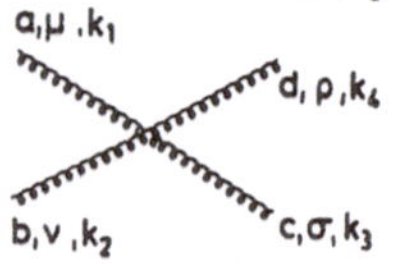

$= -i\,g^2\,(\,f^{abe}\,f^{cde} \cdot (g_{\mu\sigma}g_{\nu\rho} - g_{\mu\rho}\,g_{\nu\sigma}) + \text{symmetr})\cdot(2\pi)^4\,\delta(k_1 + k_2 + k_3 + k_4)$

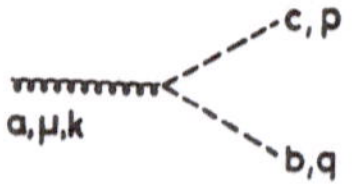

$= g\, f^{abc}\, p_\mu\,(2\pi)^4\,\delta(k + p + q)$

(b)

Fig.1.2. a) Fundamental vertex in QED. b) Fundamental vertices in QCD

6

fundamental vertex in QED for comparison. This figure shows that the structure of interactions in QCD is much richer than in QED. Because of the non-abelian nature of the gauge interaction even the theory without fermions has interaction, the three-gluon (ggg) and the four-gluon coupling (gggg). The quark-gluon coupling ($\bar{q}qg$) is similar to that of QED. It contains in addition only the colour matrix $T^a = \lambda^a/2$. Given this coupling by gT^a, gauge invariance requires the ggg coupling to be proportional to the commutator $g[T^a,T^b]$ and it is related to gf^{abc}. It is essential to recognize that all vertices contain the same coupling constant g. With the propagators (1.2.8) and the vertices in Fig.1.2 all Feynman diagrams of interest can be calculated.

The vertex $\bar{q}T^a\gamma^\mu A_\mu^a q$ is renormalized with the renormalization constant Z_1 such that the renormalized coupling g_r is

$$g_r = Z_2 Z_3^{1/2} Z_1^{-1} g \quad . \tag{1.2.9}$$

This relation will be used later in order to obtain the renormalized coupling in a specific renormalization scheme, the so-called minimal subtraction scheme. Since we have also other couplings the renormalized coupling g_r can be defined also either with the three-gluon vertex or with the ghost-ghost-gluon vertex.

By writing the renormalization constants Z_i in the form $Z_i = 1 + (Z_i - 1)$ the terms in (1.2.7) without interaction are isolated. The terms proportional to $(Z_i - 1)$ determine the subtraction terms which cancel the ultraviolet divergent parts of the Feynman diagrams. For reviews on the renormalization of gauge theories we recommend Taylor /1976/, Zinn-Justin /1975/ and Lee /1976/.

It is generally assumed that QCD is responsible for the strong force which binds quarks and gluons in the hadrons. This must be a very strong interaction, so strong that quarks and gluons are confined in the hadronic bag. However, in deep inelastic scattering, these quarks appear as freely moving, almost non-interacting particles with a coupling which is effectively small. How this feature of QCD arises will be discussed in the next section. It is clear that only for this small coupling regime we can expect that perturbation theory is applicable.

1.3 The Coupling at High Energies

If the QCD Lagrangian (1.2.2) is evaluated in perturbation theory, i.e. by a power series expansion in g, it describes a world of coloured quarks and gluons with free quarks and gluons at $t \rightarrow \pm\infty$. Since free quarks and gluons are not observed in nature, i.e. they are always confined in colour singlet hadron states, this perturbative evaluation cannot be totally realistic. On the other hand the experiments with high energy lepton beams show that the virtual quanta with large negative q^2 are scattered on quasi-free quarks and gluons which exist inside the nucleon. In the framework of QCD the scattering of the virtual photon etc. on an almost free quark is interpreted as the zeroth order approximation (g^0) in the quark-gluon coupling constant. The next higher order in g leads to the emission of an additional gluon or to the scattering of an almost free gluon with the production of a $q\bar{q}$ pair (see Fig.1.3). How should we interpret this perturbation theory in g knowing that the coupling of quarks and gluons is so large that it produces confinement? At this point we remember that g is not uniquely defined. As we explained in the last section the coupling g must be renormalized. In a theory of massless quarks — this is the appropriate approximation for high energy processes — an arbitrary mass parameter μ appears in the definition of the renormalized coupling g_r (this point will be considered in more detail in Chap.3). This parameter μ can be chosen such that the perturbation series, for example for the process in Fig.1.3, converges best. In deep inelastic scattering this parameter is chosen $\mu^2 = q^2$, where q^2 is the squared momentum transfer. Of course this makes sense only if $g^2/4\pi = \alpha_s(q^2)$ is sufficiently small. In QCD this is the case for large enough q^2 which we shall discuss next.

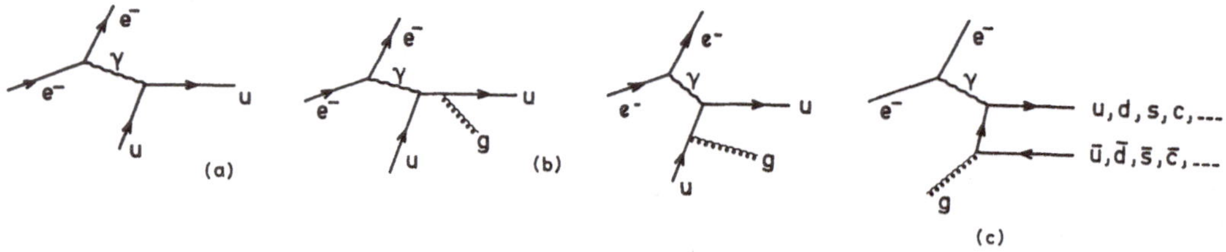

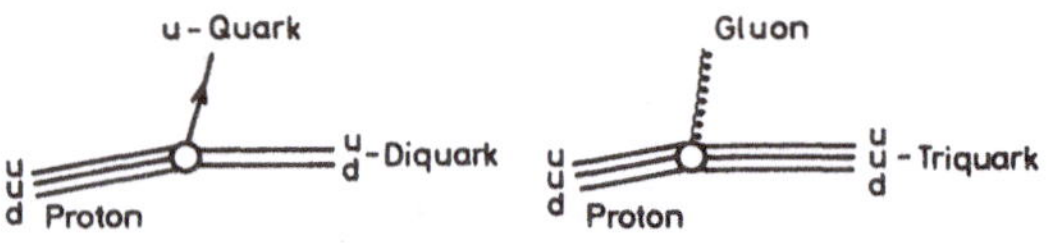

Fig.1.3. Parton model diagrams for the basic processes: a) $e^- + u \rightarrow e^- + u$, b) $e^- + u \rightarrow e^- + u + g$ and c) $e^- + g \rightarrow e^- + u\bar{u}(d\bar{d} + ...)$ together with the u quark and gluon structure function of the proton

In quantum electrodynamics we are used to consider the fine structure constant $\alpha = e^2/4\pi$ as a given fixed constant. Of course this has its origin in the fact that in all calculations the same definition, i.e. the same renormalization, of the coupling e is employed. e is defined by the electron-electron-photon vertex with all three particles on the mass shall $p_1^2 = p_2^2 = m_e^2$ and $q^2 = (p_1-p_2)^2 = 0$ (see Fig.1.2a). This is only one of many possibilities to define the renormalized charge. Any other values for the momenta p_1, p_2 or q could be chosen. Suppose we are interested to work with the coupling α which is defined for $p_1^2 = p_2^2 = m_e^2$ but for arbitrary $q^2 \neq 0$. This coupling $\alpha(q^2)$ is related to the usual coupling α, which we denote $\alpha_0 \equiv \alpha(0)$, by the following expression — considering only the lowest order term in an expansion in α_0 and assuming $q^2 \gg m_e^2$:

$$\alpha(q^2) = \alpha_0 \left(1 + \frac{\alpha_0}{3\pi} \ln \frac{q^2}{m_e^2}\right) \quad . \tag{1.3.1}$$

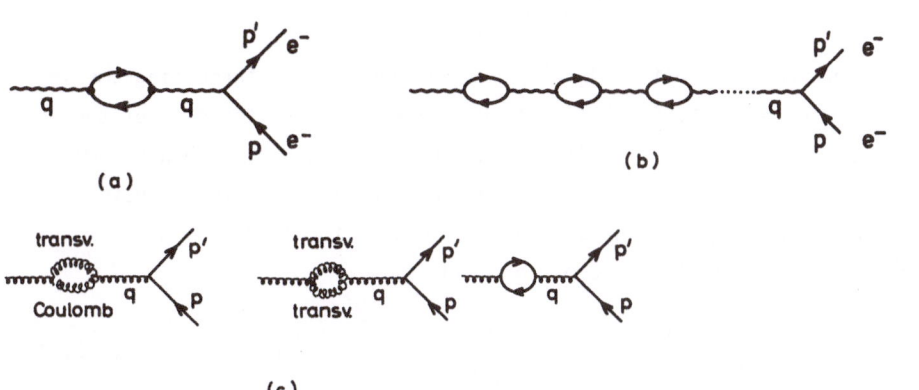

Fig.1.4. a) One-loop contribution to the photon-electron coupling in QED, b) multiloop contribution to the coupling in QED and c) one-loop contribution to the quarkgluon coupling in QCD. "Coulomb" and "transv." denote gluons with this polarization in the Coulomb gauge

This relation is obtained from the vacuum polarization contribution to the photon propagator in Fig.1.4a. Calculating also the higher order terms in α_0, shown in Fig.1.4b, in the leading logarithm approximation, we obtain terms proportional to $[\alpha_0 \ln(q^2/m_e^2)]^n$ which can be summed up with the result

$$\alpha(q^2) = \frac{\alpha_0}{1 - \frac{\alpha_0}{3\pi} \ln \frac{q^2}{m_e^2}} \quad . \tag{1.3.2}$$

In QED the summation of the series in the form (1.3.2) is not essential since $\alpha_0 = 1/137$ is very small so that even for very large q^2 the first few terms of the series in α_0 are sufficient which, of course, are taken into account in the higher order radiative corrections. For $(\alpha_0/3\pi) \ln(q^2/m_e^2) \simeq 1$ the approximations used to derive (1.3.2) break down. Therefore no statement about the behaviour of $\alpha(q^2)$ for $q^2 \to \infty$ can be made.

In QCD the behaviour of the renormalized coupling constant $\alpha_s = g^2/4\pi$ as a function of $q^2 = (p_1 - p_2)^2$ is completely different. The reason is the additional interactions of the gluon which are absent in QED. Suppose the QCD coupling has been de-

fined at the scale μ, where the renormalization has been performed. Then the relation between $\alpha_s(q^2)$, the coupling at scale $\sqrt{q^2}$, is calculated from the diagrams in
Fig.1.4c. In this calculation a physical gauge, the Coulomb gauge, is used, otherwise the structure of the diagrams would be more complicated. Up to order g^2 the
relation between $\alpha_s(q^2)$ and $\alpha_s(\mu^2)$ is

$$\alpha_s(q^2) = \alpha_s(\mu^2) \left[1 + \frac{\alpha_s(\mu^2)}{4\pi} \left(5 - 16 + \frac{2N_f}{3} \right) \ln\frac{q^2}{\mu^2} \right] \quad . \tag{1.3.3}$$

The terms in the bracket in front of $\ln(q^2/\mu^2)$ correspond to the three diagrams in
Fig.1.4c. For $N_f < 16$, where N_f is the number of quark flavours, the sign of the
factor of $\ln(q^2/\mu^2)$ is negative, opposite to the QED case. This sign change comes
from the second diagram in Fig.1.4c, the 2-gluon contribution with two transversal
gluons. Then summing up all higher order contributions in the leading logarithm approximation yields the result (N_c = number of colours)

$$\alpha_s(q^2) = \alpha_s(\mu^2) \left[1 + \frac{\alpha_s(\mu^2)}{4\pi} \left(\frac{11}{3}N_c - \frac{2}{3}N_f \right) \ln\frac{q^2}{\Lambda^2} \right]^{-1} \quad . \tag{1.3.4}$$

Since the coefficient of $\ln(q^2/\mu^2)$ is positive, the limit for $q^2 \to \infty$ can be taken
with the well known result $\alpha_s(q^2) \to 0$. This means, that if in a process a scale q^2
appears which is sufficiently large, then the coupling constant α_s is small. This
property of QCD is called asymptotic freedom and was derived the first time by Gross
and Wilczek /1973/ and by Politzer /1973/. For a more recent discussion see Hughes
/1980/. This property is very important for the interpretation of QCD perturbation
theory. It justifies the assumption that for high enough energies perturbation theory may be sufficiently convergent. How large q^2 should be depends on the proper
scale, i.e. on the value of α_s for a particular $q^2 = \mu^2$ which must be determined
by experiment. Furthermore only after higher order corrections for a process have
been computed, it is known how the kinematic variables of the process are related
to the effective q^2 which make the higher order corrections small. This will be
made more transparent later when we discuss higher order QCD corrections for jet
cross sections.

A measure for the scale of α_s is defined by the formula

$$\alpha_s(q^2) = \frac{4\pi}{\left(\frac{11}{3}N_c - \frac{2}{3}N_f \right) \ln\frac{q^2}{\Lambda^2}} \quad . \tag{1.3.5}$$

This formula contains the same information as (1.3.4) concerning the q^2 dependence
of α_s. Only the boundary conditions are different, in (1.3.5) $\alpha_s = \infty$ for $q^2 = \Lambda^2$
whereas in (1.3.4) $\alpha_s = \alpha_s(\mu^2)$ for $q^2 = \mu^2$. The parameter Λ gives us the measure

for the value of α_s. Of course (1.3.5) is useful only for $q^2 \gg \Lambda^2$ since non-leading terms in the leading logarithm approximation are neglected. Later we shall see that $\Lambda \simeq 0.5$ GeV, so that α_s becomes small for $q^2 \gg \Lambda^2 \simeq 0.25$ GeV^2.

Using $\alpha_s(q^2)$ instead of $\alpha_s(\mu^2)$ as perturbative expansion parameter has the effect that specific contributions are summed up to arbitrary order in $\alpha_s(\mu^2)$.

In the limit of infinite q^2 the gluon corrections to the parton model (see Fig. 1.3) vanish. Therefore in this limit the results of perturbative QCD approach the naive parton model. But we must keep in mind that $\alpha_s(q^2)$ is effective only for the elementary scattering process in terms of quarks and gluons and not for the transition processes quark or gluons ↔ hadrons (see Fig.1.3). The idea therefore is, that only this elementary scattering process can be calculated in QCD perturbation theory and the decay functions of quarks and gluons or the structure functions of hadrons must be supplied by other means. The calculation of such transition amplitudes hadron ↔ quark or gluon can be done non-perturbatively only since it belongs to the confinement region with small effective q^2, i.e. large α_s.

It is important to realize that perturbative calculations always involve this separation of non-calculable — at least in perturbative theory — long-distance (or low energy) effects from the calculable short distance (or high energy) contributions. As we shall see later the mechanism for the production of jets in e^+e^- annihilation are calculable in QCD perturbation theory but not the details of hadrons emerging from these jets.

Before we go over to the study of jets in the next chapter we shall collect some characteristic features of the QCD Lagrangian which should be borne out by the analysis of high energy jet phenomena:

(i)   quarks with spin 1/2 exist as colour triplets

(ii)   gluons with spin 1 exist as colour octets

(iii)   the coupling $q\bar{q}g$ exists

(iv)   the couplings ggg and $gggg$ exist

(v)   the coupling constants in (iii) and (iv) are equal

(vi)   the effective universal coupling g decreases with increasing energy scale like $\ln q^2$.

2. Electron-Positron Annihilation into Hadron Jets

2.1 e^+e^- Annihilation in the Parton Model

The notion of jets in e^+e^- annihilation is closely connected with the discovery of Bjorken scaling in deep inelastic electron-nucleon scattering in 1968 at SLAC. As was already mentioned in the introduction, the experiments revealed that the inelastic electron scattering proceeds in such a way as if the spacelike ($q^2 < 0$) virtual photon interacts with pointlike constituents of the nucleon, the partons, now identified with u and d quarks inside the proton or neutron /Feynman, 1972/.

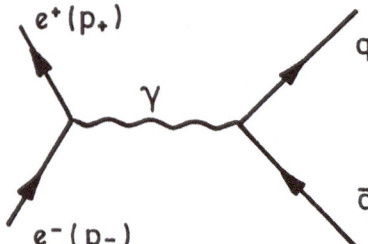

Fig.2.1. Diagram for $e^+e^- \to q\bar{q}$ describing hadron production in the parton model and zeroth order QCD contribution

The equivalent process with a timelike ($q^2 > 0$) virtual photon is the e^+e^- annihilation into a quark-antiquark pair: $e^+e^- \to q\bar{q}$, as shown in Fig.2.1. In this simple model the virtual photon, produced by the annihilating leptons, creates a quasi-free quark and antiquark, which have the tendency to move with opposite directed but equal momenta away from each other. This is prevented by the confinement forces, so that at much later times ($t \simeq 1\ \text{GeV}^{-1}$) the quark and antiquark transform with unit probability into hadrons. These hadrons should come out roughly with the momentum direction of the original quark and antiquark, so that two distinct particle jets with opposite directed momenta evolve. This simple picture /Drell, Levy, Yan, 1970; Cabibbo, Parisi, Testa, 1970; Berman, Bjorken, Kogut, 1971/ had its early support from the fact that the total annihilation cross section for hadron production is given by the square of the quark charges Q_f

$$\sigma(e^+e^- \to q\bar{q} \to \text{hadrons}) = \frac{4\pi\alpha^2}{3W^2} \, N_c \sum_f Q_f^2 \tag{2.1.1}$$

where the sum 'f' is over all active flavours which come in N_c colours. Dividing by $\sigma(e^+e^- \to \mu^+\mu^-)$ we obtain the famous ratio R

$$R = \frac{\sigma(e^+e^- \to q\bar{q})}{\sigma(e^+e^- \to \mu^+\mu^-)} = N_c \sum_f Q_f^2 = \begin{cases} 2 & \text{for } f = u,d,s \\[2mm] \dfrac{10}{3} & \text{for } f = u,d,s,c \\[2mm] \dfrac{11}{3} & \text{for } f = u,d,s,c,b \\[2mm] \dfrac{16}{3} & \text{for } f = u,d,s,c,b,t \quad . \end{cases} \tag{2.1.2}$$

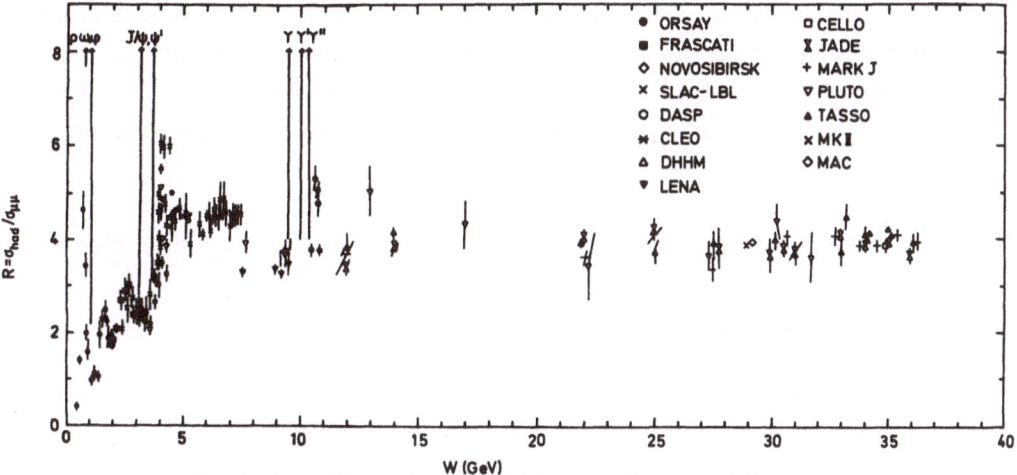

Fig.2.2. The ratio of the total cross section for e^+e^- annihilation into hadrons to $\mu^+\mu^-$ production cross section as a function of the total e^+e^- energy W

A recent compilation of all measured R values is shown in Fig.2.2 where the transitions from regions where the (u,d,s) quarks are excited to the (u,d,s,c) and to the (u,d,s,c,b) regions are clearly visible. The largest deviation from the values in (2.1.2) occurs above the charm threshold, which might be due to systematic measurement errors. A recent R value for $7.4 \leq W \leq 9.4$ GeV is $R = 3.37 \pm 0.06 \pm 0.23$ /Niczyporuk et al., 1982/ in good agreement with 10/3. An average R value for PETRA energies (30 < W < 36.8 GeV) is /Wolf, 1982/:

$$\bar{R}(W = 34 \text{ GeV}) = 3.96 \pm 0.10 \quad . \tag{2.1.3}$$

When comparing this $\bar{R}$ with (2.1.2) we must take into account that for large W the neutral currents of the weak interaction, i.e. the Z exchange, also contribute to $e^+e^- \to q\bar{q}$, which is not contained in (2.1.2). At W = 34 GeV this amounts to $\Delta R = 0.03$ if $\sin^2\theta_W = 0.23$, which must be subtracted from (2.1.3) if we want to compare with (2.1.2) /Wolf, 1982/.

We see that the experimental results for R are remarkably consistent with (2.1.2). This not only shows the consistency with the standard charge assignment, but also reveals the number N_c of colours is equal to 3.

Later we shall see that this simple result of the naive quark-parton model still holds in quantum chromodynamics in the limit $W^2 \to \infty$. Therefore it provides not only a test of this underlying theory of the quark-gluon interaction but it serves also as a demonstration of renormalization-group-improved perturbation theory /Appelquist and Georgi, 1973; Zee, 1973; Politzer, 1974/.

Other presumed properties of the simple parton model are: (i) The produced hadrons have a limited transverse momentum p_T relative to the direction of momentum of the originally produced quarks. This means, the produced hadrons come out as jets. (ii) The inclusive cross section as a function of the hadron momentum p scales according to

$$\frac{d\sigma}{dx} = \frac{8\pi\alpha^2}{W^2} \sum_f Q_f^2 D_f^h(x) \tag{2.1.4}$$

where (in the e^+e^- center-of-mass frame $\mathbf{q} = 0$)

$$x = p/p_{quark} = 2p/W \tag{2.1.5}$$

is the scaled momentum. This property is analogous to scaling of the structure functions $F_1(x)$ and $F_2(x)$ in deep inelastic lepton scattering. The function $D_f^h(x)$ measures the probability of finding a hadron of type h emerging from a quark with flavour f. A comparison of single charged hadron spectra for energies up to 34 GeV can be seen in Fig.2.3 /Felst, 1981/. In the comparison we notice deviations from the scaling property (2.1.4). This means that $D_f^h(x)$ depends also on $W^2 \equiv q^2$. This scaling violation should be seen in analogy to the scaling violation in the structure functions $F_1(x,q^2)$ and $F_2(x,q^2)$ observed in inelastic lepton-nucleon scattering. Of course the scaling (2.1.4) or its weak violation is to be expected only towards the boundaries of phase space, i.e. for larger x values. This is the quark fragmentation region in contrast to the so-called current fragmentation region near $y \simeq 0$, where y is the rapidity variable. This is defined with respect to the momentum direction of the originally produced quark-antiquark pair giving the longitudinal momentum from which y is calculated. We expect the rapidity distribution of produced non-leading

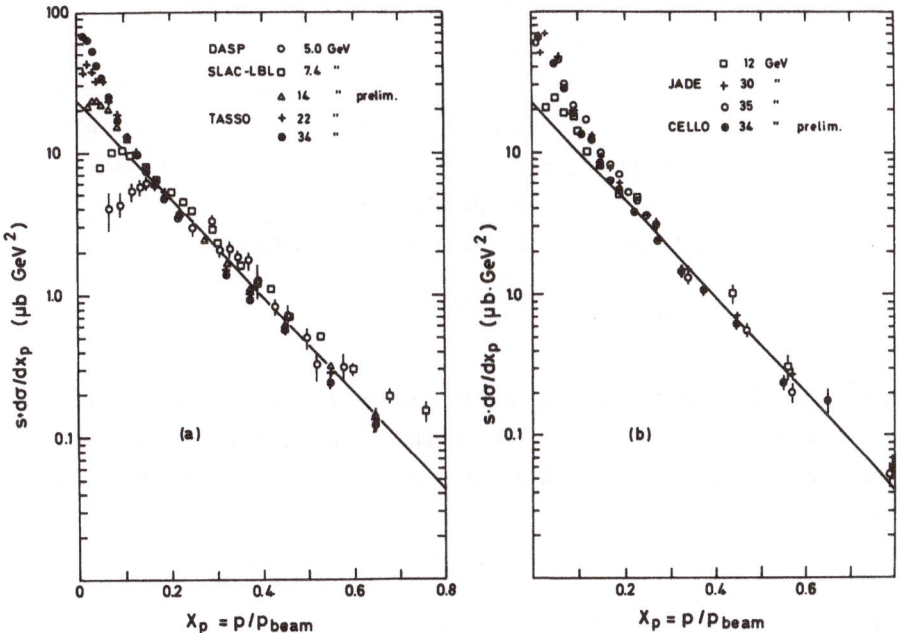

Fig.2.3. One-particle distribution $s \cdot d\sigma/dx_p$ for charged hadrons for c.m. energies $W \equiv \sqrt{s}$ from 5 to 34 GeV. a) Shows data from TASSO at PETRA, SLAC-LBL at SPEAR and DASP at DORIS. b) Shows data from CELLO and JADE. The full lines are drawn to guide the eye and correspond to $s \cdot d\sigma/dx_p = 23 \exp(-8x_p)\mu b$ /Felst, 1981/

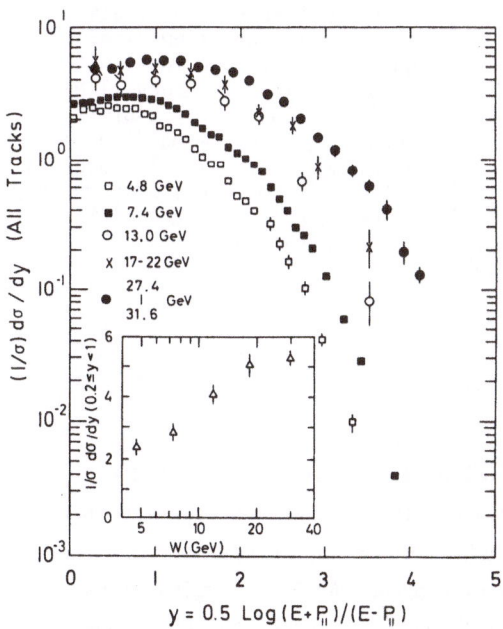

Fig.2.4. Rapidity spectrum $(1/\sigma)d\sigma/dy$ for charged hadrons (pion mass assumed) measured by SLAC-LBL at 4.8 and 7.4 GeV /Hanson et al., 1982/ and by TASSO between 13 and 31.6 GeV /Brandelik et al., 1980d/. y is measured with respect to the sphericity axis (SLAC-LBL) or thrust axis (TASSO). The insert shows $(1/\sigma)d\sigma/dy$ averaged over $0.2 < y < 1$ as a function of c.m. energy W

15

hadrons to be uniform, just as in ordinary hadron reactions. This is shown in Fig. 2.4 /Wolf, 1981/. From this, i.e. no increase of $(1/\sigma)d\sigma/dy$ near $y \simeq 0$ with increasing W, it would follow, that the multiplicity would grow only logarithmically with W as in ordinary hadron collisions at low energies. Figure 2.4 shows clearly that this is not the case. The plateau near $y \simeq 0$ rises with increasing energy. Therefore the multiplicity must rise stronger than $\ln q^2$. Recent analysis shows that the average multiplicity of charged particles in e^+e^- annihilation behaves like

$$\langle n_{ch} \rangle = 2 + 0.2 \ln q^2 + 0.18 (\ln q^2)^2 \quad . \tag{2.1.6}$$

The dependence of the multiplicity $\langle n_{ch} \rangle$ as a function of $W = \sqrt{s}$ is shown in Fig.2.5 compared with some other model predictions (see figure caption) /Wolf, 1981/. But the stronger increase than $a + b \ln q^2$ is clearly seen in agreement with the increase of the rapidity plateau near $y \simeq 0$.

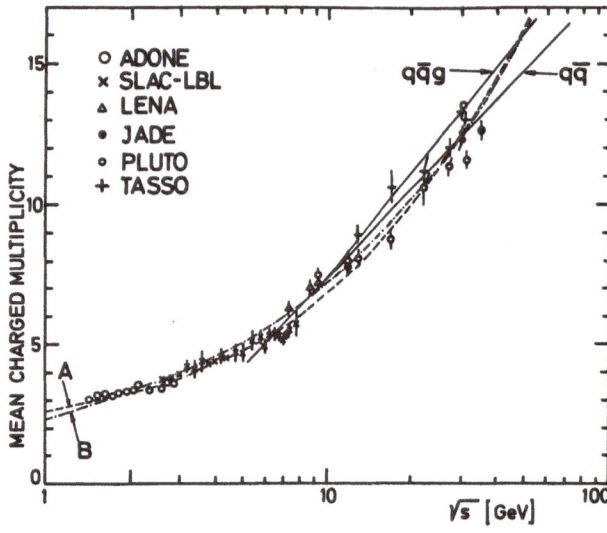

Fig.2.5. Mean charged multiplicity as a function of c.m. energy $W \equiv \sqrt{s}$ compared to various model predictions: straight line labelled $q\bar{q}$ is for $q\bar{q}$ model with subsequent fragmentation, straight line labelled $q\bar{q}g$ is for $q\bar{q} + q\bar{q}g$ model with fragmentation. The curve labelled (A) is the best fit of the form $\langle n_{ch} \rangle = n_0 + a \exp[b\sqrt{\ln(s/\Lambda^2)}]$ predicted in leading log approximation to QCD. Curve B shows the dependence $\langle n_{ch} \rangle = 2.3\, s^{1/4}$ (s in GeV^2)

2.2 First Experimental Evidence for Jets

Although the production of jets in e^+e^- annihilation as a manifestation of the process $e^+e^- \rightarrow q\bar{q}$ was suggested by Bjorken and Brodsky /1970/ it was not earlier than 1975 that they were discovered experimentally, when high enough center-of-mass energies W became available at SLAC's e^+e^- storage ring SPEAR. At low energies it was not possible to see jets because the two jet cones were too broad. Let us suppose that the transverse momentum p_T with respect to the jet direction (which, theoretically, is the momentum direction of the primordial quark) is limited and that the multiplicity grows only logarithmically as in (2.1.6) the jet cone becomes narrower and narrower with increasing energy W. Let $<n>$ be the average particle multiplicity, $<p_T>$ and

$$<p_{||}> \simeq <p> \simeq \frac{W}{<n>} \qquad (2.2.1)$$

the average transverse and longitudinal hadron momenta, then the mean half angle $<\delta>$ of the jet cone (see Fig.2.6)

$$<\delta> \simeq \frac{<p_T>}{<p>} \simeq \frac{<p_T> <n>}{W} \sim \frac{1}{W} \quad . \qquad (2.2.2)$$

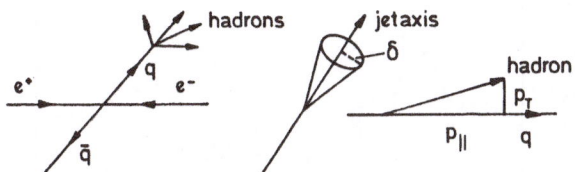

Fig.2.6. Hadron emission by quark and definition of jet cone

This means that the jet cone opening angle decreases with increasing total e^+e^- energy W like 1/W. Thus, for example, for W = 4 GeV the measured charged multiplicity is roughly 4 (see Fig.2.5), giving a total multiplicity of nearly 6, so that with $<p_T>$ = 0.32 GeV we get

$$<\delta> \simeq 0.48 \simeq 30° \quad . \qquad (2.2.3)$$

This value is presumably an underestimate for the actual situation. It shows that even at W = 4 GeV each of the two jet is broader than 60° in average.

Jets were first seen in e^+e^- annihilation experiments at SPEAR with the MARK II detector of the SLAC-LBL Collaboration /Hanson et al., 1975,1982/. In order to establish the jets it is necessary to prove the limited transverse momentum with respect to a jet axis. The jet must be determined from the momenta of the produced hadrons. In this early work the jet axis was defined in terms of sphericity introduced by Bjorken and Brodsky /1970/ which is

$$S = \frac{3}{2} \min \frac{\sum_i |\mathbf{p}_{iT}|^2}{\sum_i |\mathbf{p}_i|^2} \quad . \qquad (2.2.4)$$

In (2.2.4) the $\mathbf{p}_{iT}$ are the transverse particle momenta of all particles in an event relative to an axis which is chosen in such a way that the numerator in (2.2.4) is minimal. Comparing (2.2.4) with (2.2.2) we see that the sphericity S is roughly the average of the square of the jet cone opening angle:

$$\langle S \rangle \simeq \frac{3}{2} \langle \delta^2 \rangle \simeq \frac{3}{2} \frac{\langle p_T^2 \rangle \langle n \rangle^2}{W^2} \quad . \tag{2.2.5}$$

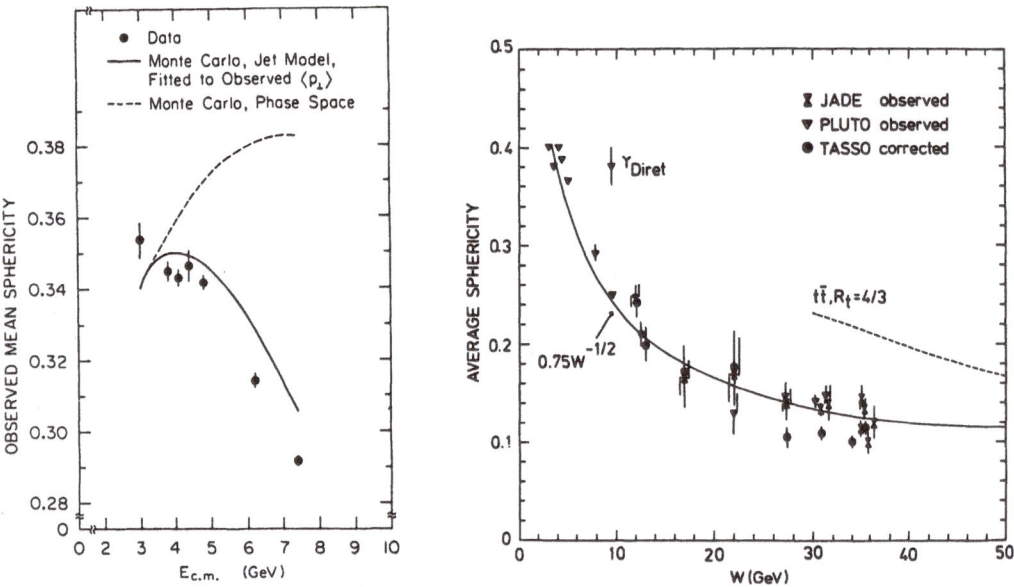

Fig.2.7. Observed mean sphericity vs $E_{c.m.} \equiv W$ for data, jet model (solid curve) and phase space model (dashed curve). /Hanson et al., 1982/

Fig.2.8. Average sphericity as a function of c.m. energy W as measured by JADE, PLUTO and TASSO at PETRA. Dashed curve is model prediction for $e^+e^- \to t\bar{t}$ with 2/3 charged t quark

The first experimental data for the mean sphericity are shown in Fig.2.7. They come from the SLAC-LBL group /Hanson et al., 1975,1980/ and represent the first evidence for jet formation in e^+e^- annihilation. The mean sphericity is roughly constant as a function of the total e^+e^- energy $E_{c.m.} \equiv W$ up to 4 GeV and then decreases with increasing $E_{c.m.}$. The solid and dashed curves show the Monte Carlo results for a jet model with $\langle p_T \rangle$ = 315 MeV and for a simple phase-space model. At lower energies (≤ 4 GeV), where $\langle p_\| \rangle$ is of the same order of magnitude as $\langle p_T \rangle$ both models predict the same average sphericity. Above 4 GeV the jet model describes the data very well whereas in the phase space model the sphericity rises with energy in disagreement with the data. In Fig.2.8 we show a recent compilation of experimental data for $\langle S \rangle$

for energies up to 36 GeV /Wolf, 1982/. We see how <S> decreases with increasing W. The decrease of <S> is, however, much less than predicted by (2.2.5), which is <S> ~ $1/W^2$. The data decrease more like <S> ~ $1/W^{1/2}$. Since <S> is a measure of the square of the half jet opening angle δ introduced in (2.2.2) (<S> = 3/2 <δ^2>) [see (2.2.5)] according to Fig.2.8 <δ> decreases from <δ> ≃ 30° at W = 5 GeV down to <δ> ≃ 17° at W = 35 GeV. That <δ> does not decrease faster with increasing W apparently comes from the strong increase of the average multiplicity <n> as can be seen in Fig.2.5, which diminishes the decrease of <δ^2> because of (2.2.5).

In order to discover the two back-to-back jets in the final hadron state at these still relatively low energies of the SPEAR storage ring, it was essential to perform the sphericity analysis and to study <S> as a function of the beam energy. If, instead, one studies $d\sigma/dp_T$ at an already large, but fixed energy, the p_T distributions for the jet model and the phase space model do not differ very much. This is seen in Fig.2.9, where this comparison is shown for W = 7.4 GeV together with the data, which, of course, are well described by the model with a limited transverse momentum p_T.

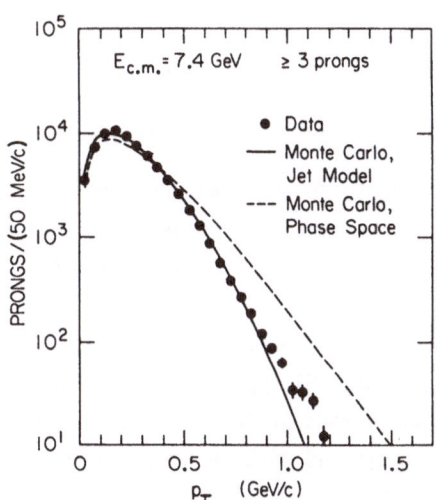

Fig.2.9. Observed single-particle p_T with respect to jet axis with three or more detected charged particles for 7.4 GeV data compared with jet model (solid curve) and phase space-model (dashed curve) predictions /Hanson et al., 1982/

A similar comparison is shown in Fig.2.10. Here the measured sphericity distribution $d\sigma/dS$ at 3.0, 6.2 and 7.4 GeV is compared to calculated distributions of the jet model and the phase-space model. At 3.0 GeV the data agree with the predictions of either the phase-space or the jet model. At this energy the limiting of transverse momentum to an average of 0.35 GeV has no effect on the phase-space particle distributions since transverse and longitudinal momentum are of the same order and do not exceed 0.35 GeV to a large extent. At 6.2 and 7.4 GeV the S distributions are

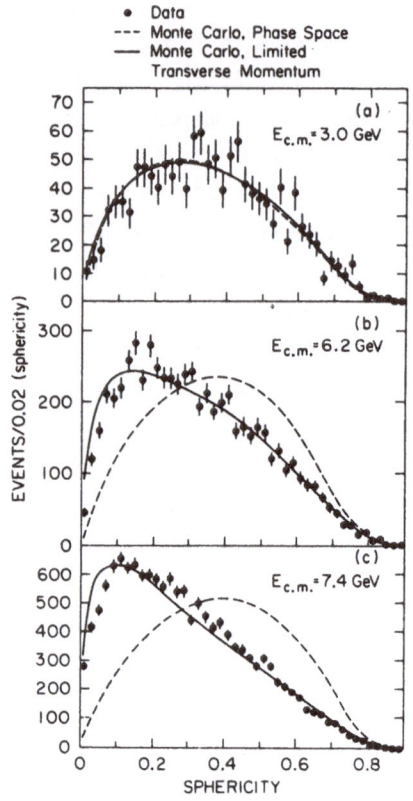

- Data
- --- Monte Carlo, Phase Space
- —— Monte Carlo, Limited Transverse Momentum

EVENTS/0.02 (sphericity)

(a) $E_{c.m.} = 3.0$ GeV

(b) $E_{c.m.} = 6.2$ GeV

(c) $E_{c.m.} = 7.4$ GeV

SPHERICITY

Fig.2.10. Observed sphericity distributions for data, jet model (solid curve) and phase-space model (dashed curve) for a) $E_{c.m.} = 3.0$ GeV, b) $E_{c.m.} = 6.2$ GeV and $E_{c.m.} = 7.4$ GeV

peaked toward low S favouring the jet model over the phase space model. But the distributions in S are still very broad and to establish the evidence for jet structure in e^+e^- hadron production is only possible via direct comparison with model predictions.

These findings were confirmed later with the DORIS ring at DESY with somewhat higher energies near the Υ resonance /Berger et al., 1981a; Niczyporuk et al., 1981/.

This is completely different at PETRA energies, where most of the events are two jet and which, because of the narrower cones, can be selected with the eye. An example of such an event which was produced at W = 31.6 GeV is reproduced in Fig.2.11. The small transverse momenta with respect to the jet axis are easily visible.

Other tests of the quark-parton model, as for example, the scaling behaviour of single hadron spectra are also positive and had been shown already in Fig.2.3.

Another very important test of the underlying quark structure of the jet in e^+e^- annihilation is the measurement of the angular distribution $d\sigma/d\cos\theta$ of the jet axis with respect to the beam direction. It is well known that the distribution for the production of massless spin 1/2 particles is (see for example /Gatto, 1965/)

$$d\sigma/d\cos\theta \sim 1 + \cos^2\theta \quad .$$ (2.2.6)

20

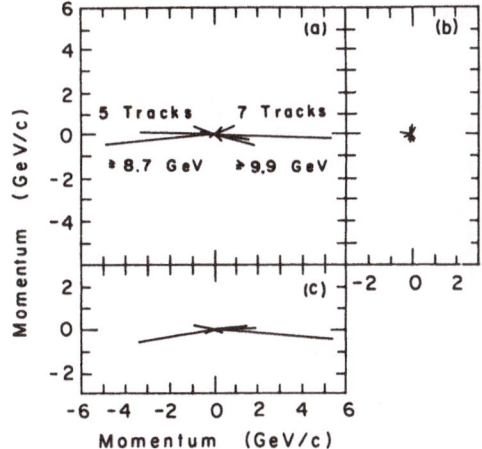

Fig.2.11. Momentum-space drawing of two-jet event in each of three projections: a) = n_2-n_3 plane, b) = n_1-n_2 plane, c) = n_1-n_3 plane

The formula (2.2.6) applies also to $e^+e^- \to q\bar{q}$ for massless quarks (a valid assumption at least for u, d, s quarks) and would manifest itself as the angular distribution of the jet axis, i.e. the sphericity axis, with respect to the beam axis.

All experimental measurements of jet axis angular distributions so far are consistent with (2.2.6). The first data came from the SLAC-LBL Collaboration at SPEAR. They used transversal polarized e^+ and e^- beams. This makes the determination of the θ distribution more feasable at these low e^+e^- energies. With transverse polarized beams the angular distribution has the following form

$$d\sigma/d\Omega \sim 1 + \alpha\cos^2\theta + \alpha P_+P_-\sin^2\theta\cos2\phi \quad . \tag{2.2.7}$$

In (2.2.7) φ is the azimuthal angle of the jet axis with respect to the storage ring plane and P_+ and P_- are the polarizations of the e^+ and e^- beams, respectively. The measured φ distributions (averaged over θ) for 6.2 and 7.4 GeV are shown in Fig. 2.12. At 6.2 GeV the beam polarization P_+ = P_- = 0 and therefore an isotropic φ distribution is observed whereas at 7.4 GeV where P_+P_- = 0.5 they observed the characteristic cos2φ behaviour. From this measurement the value α = 0.97 ±0.14 has been deduced in agreement with the expectation (2.2.6). Similar results, but with less accuracy were obtained by the PLUTO Collaboration at DORIS for W = 7.7 GeV and 9.4 GeV /Berger et al., 1978/.

In Fig.2.13 a more recent measurement of the θ distribution at higher energies is shown. The results come from the JADE Collaboration at PETRA and are for energies between 30 and 35 GeV. In this case the e^+ and e^- beams were unpolarized. Thus, although the beam energies were much higher, so that the jet axis is much better defined, the value of α /Elsen, 1981/

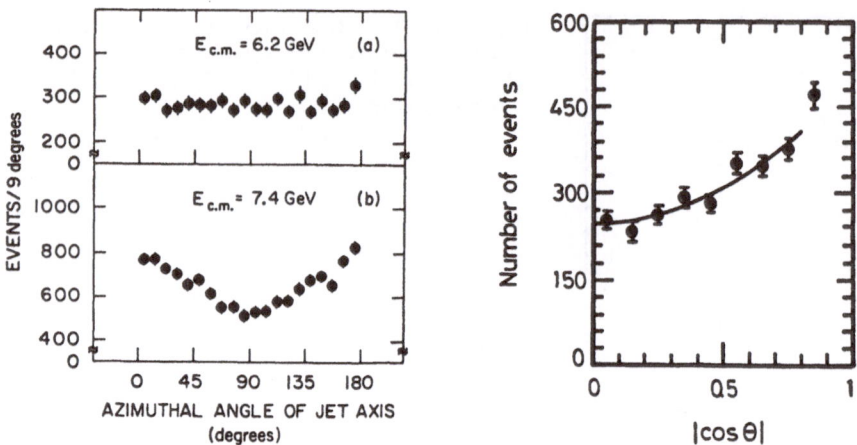

Fig.2.12. Observed distribution of jet-axis azimuthal angles ϕ from the plane of the storage ring for jet axis with /cosθ/ $\leq$ 0.6 for a) $E_{c.m.}$ = 6.2 GeV and b) $E_{c.m.}$ = 7.4 GeV /Hanson et al., 1982/

Fig.2.13. Angular distribution of sphericity axis with respect to beam direction of the form $1 + 1.04 \cos^2\theta$ fitted to JADE data /Elsen, 1981/

$$\alpha = 1.04 \pm 0.16 \qquad\qquad\qquad\qquad\qquad\qquad (2.2.8)$$

does not have a better accuracy than the SPEAR value. The jet axis was also defined with the sphericity. We see that (2.2.8) is also consistent with (2.2.6).

This test of the spin 1/2 nature of the-quarks produced in e^+e^- annihilation is analogous to the verification of the Callan-Gross relation /Callan, Gross, 1969/ in deep inelastic lepton-nucleon scattering: $F_2(x) = 2xF_1(x)$ which is also very well satisfied experimentally.

As a last point we may ask the question to what extent do all these tests verify the existence of jets in e^+e^- annihilation. It is clear that there exist also models with other input which would describe the data. For example, the $(1 + \cos^2\theta)$ angular distribution could be explained via the production of two resonances with opposite normality (= parity $\cdot$ (-1)^S, where S is the spin of the resonance), like $\pi\omega$, πA_2 etc. /Kramer and Walsh, 1973/. But in such a model, with increasing energy, always new resonances with the same property must appear which is rather unlikely. Actually in the data of the SLAC-LBL experiment no evidence for any structure in the mass distribution for three-prong, charge = $\pm$ 1 jets, was found. Therefore we must conclude that there is no evidence for copious production of resonances which could lead to jet structure in the majority of the events.

We conclude that the results of the SLAC-LBL experiment and the experiments at DORIS are in general agreement with the predictions of the quark-parton model and it seems that no other model of sufficient simplicity exists which could explain these data as well.

It is clear that with the higher PETRA energies many more tests of the $q\bar{q}$ production mechanism have been performed. We mention in this respect in particular the work on the existence of a long range charge correlation between the quark fragmentation regions of the two jets /Brandelik et al., 1981; see also Peterson, 1980/.

For large enough energies new effects appear through interference of photon and Z exchange. This modifies the angular distribution by adding a term proportional to $\cos\theta$ whose strength depends on the charge of the produced quark. A recent reference is Schiller /1979/.

2.3 On Jet Measures

In Sect.2.2 we have seen already that it is important to have variables which measure the jettiness of an event or even more complicated measures which allow to say whether a hadronic event has 2, 3 or 4 jets. Of the first category we introduced already sphericity. In the following we shall consider some other jet variables which will appear at various places in this review. Actually sphericity appears in a more general context if we consider the sphericity tensor $S_{\alpha\beta}$ /Bjorken and Brodsky, 1970/

$$S_{\alpha\beta} = \frac{\sum_i p_{i\alpha} p_{i\beta}}{\sum_i \mathbf{p}_i^2} . \tag{2.3.1}$$

In (2.3.1) the index i runs over all particles of an event and p_α, p_β are the momentum components of a particle (α, β = 1, 2, 3). The momentum tensor $S_{\alpha\beta}$ can be transformed into diagonal form. This way one obtains the principal axes $\mathbf{n}_1$, $\mathbf{n}_2$, $\mathbf{n}_3$ with corresponding eigenvalues Q_1, Q_2 and Q_3. The Q_i are ordered, $Q_1 < Q_2 < Q_3$, and normalized so that $Q_1 + Q_2 + Q_3 = 1$. Then $\mathbf{n}_3$ is the axis concerning which of the squares of the transverse momenta are minimal. Therefore the sphericity as defined in (2.2.4) is

$$S = \frac{3}{2}(1-Q_3) = \frac{3}{2}(Q_1+Q_2) \tag{2.3.2}$$

and the sphericity axis is equal to $\mathbf{n}_3$. S = 0 for events with two particles with equal and opposite momenta (ideal two-jet event) and S goes to 1 for completely isotropic events. Since $Q_1 + Q_2 + Q_3 = 1$ only two of the eigenvalues are needed to characterize an event. For example one can take S and the so-called aplanarity Ap

$$Ap = \frac{3}{2} Q_1 = \frac{3}{2} \min \frac{\sum_i |\mathbf{p}_{iT,out}|^2}{\sum_i \mathbf{p}_i^2} \quad . \tag{2.3.3}$$

The aplanarity Ap minimizes the transversal momentum $\mathbf{p}_{T,out}$ with respect to a plane. Events with small Ap are almost planar events. $\mathbf{n}_1$ is normal to this plane, in which $\mathbf{n}_2$ and $\mathbf{n}_3$ lie. The variables of an event given by its values for Q_1, Q_2 and Q_3 can be plotted inside a triangle as shown in Fig.2.14 where events obtained by the TASSO Collaboration at PETRA are plotted. In this triangle planar events are found in the strip with small Ap values, 2-jet events have in addition also small S. The method based on the tensor (2.3.1) first applied by Alexander /1978/ (see also /Wu and Zobernig, 1979/) to the analysis of the 3-gluon decay of the Υ resonance, has the virtue that the eigenvalues Q_i and the principal axes $\mathbf{n}_i$ and from this S and Ap can be calculated quite easily for every event. However, since the momenta enter quadratically, high momentum particles enter with a stronger weight in the determination of S and Ap. Also it is not invariant against clustering (multiplicity) of particles and depends stronger on details of the fragmentation of quarks into hadrons. This

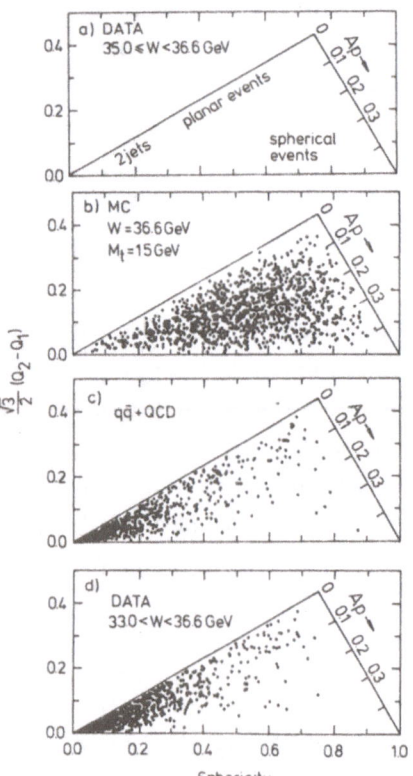

Fig.2.14. The sphericity plot a) location of events of different character, b) prediction for $e^+e^- \rightarrow t\bar{t}$ (M_t = 15 GeV), c) prediction for $q\bar{q} + q\bar{q}g$, d) data from TASSO /Brandelik et al., 1980b/

means for example, that the sphericity changes if a particle momentum splits by decay, for instance $\rho^0 \to \pi^+\pi^-$, or by fragmentation, for instance $q \to q' + \text{meson}$, into two or more momenta. Since these variables are not insensitive against clustering they are also sensitive to the emission of soft or collinear gluons. There exist, however, variables without this property. They are called infrared safe. These quantities are all made out of linear sums of momenta. The best known examples are

$$\text{Thrust} \qquad T = \max \frac{\sum\limits_i |\mathbf{p}_{iL}|}{\sum\limits_i |\mathbf{p}_i|} \qquad\qquad (2.3.4)$$

$$\text{Spherocity} \qquad S' = \left(\frac{4}{\pi}\right)^2 \min\left(\frac{\sum\limits_i |\mathbf{p}_{iT}|}{\sum\limits_i |\mathbf{p}_i|}\right)^2 \qquad\qquad (2.3.5)$$

$$\text{Acoplanarity} \qquad A = 4\min\left(\frac{\sum\limits_i |\mathbf{p}_{iT,\text{out}}|}{\sum\limits_i |\mathbf{p}_i|}\right)^2 \quad . \qquad\qquad (2.3.6)$$

For thrust, which was introduced by Brandt, Peyrou, Sosnovski, Wroblewski /1964/ and rediscovered by Farhi /1977/, the thrust axis $\mathbf{n}$ is defined in such a way, that the longitudinal momenta p_{iL} with respect to this axis are maximal. The values for T range from 0.5 to 1, where the first value pertains to isotropic events and the last one to completely collinear configurations. In a similar way, for spherocity S' the $|\mathbf{p}_{iT}|$ is minimized. It lies between 0 and 1 for configurations from collinear to completely isotropic. Similarly in the acoplanarity A the $\mathbf{p}_{iT,\text{out}}$ is measured in such a way that A is minimal with respect to a plane. Planar events must have small A values. For massless particles A lies between 0 and 2/3.

In connection with the analysis to verify the existence of 4 QCD jets in Chap.3, i.e. contributions $e^+e^- \to q\bar{q}gg$ and $e^+e^- \to q\bar{q}q\bar{q}$, we shall encounter the variable tripodity D_3, first introduced by Nachtmann and Reiter /1982b/. Tripodity is defined by

$$D_3 = 2\max\left(\frac{\sum\limits_i |\mathbf{p}_{iT}| \cos^3 \angle(\mathbf{n},\mathbf{p}_{iT})}{\sum\limits_i |\mathbf{p}_i|}\right) \quad . \qquad\qquad (2.3.7)$$

In (2.3.7) the $\mathbf{p}_{iT}$ are the projections of the particle momenta p_i on a plane perpendicular to the thrust axis and $\mathbf{n}$ is a unit vector in this plane, fixed in such a way that the quantity in brackets is maximal. Thus D_3 measures the momentum distribution of an event in a plane normal to the thrust axis. For a symmetrical momentum distribution in this plane we have $D_3 = 0$ because of the odd power of $\cos\angle(\mathbf{n},\mathbf{p}_{iT})$ in (2.3.7). Therefore for 2 and 3 parton final states $D_3 = 0$. In the 4 parton final states we have two separate classes. In the first class two partons are on both sides of the plane perpendicular to the thrust axis, respectively. Then $D_3 = 0$ for this class because of the symmetric distribution. In the other class one high energy parton is on one side of the plane and the three others come out on the other side, so that $D_3 \geq 0$ (in general $0 \leq D_3 \leq 0.325$).

Various other jet measures have been constructed. For example, a generalization of thrust to three clusters instead of two, called triplicity /Brandt, Dahmen, 1979/, or jettiness /Wu, Zobernig, 1979/. For algorithms to determine some of these quantities and the problems involved on should read Brandt, Dahmen /1979/. For an algorithm to detect four jets one should also consult Wu /1981/. See also Bopp /1979/.

2.4 Fragmentation of Quarks and Gluons

2.4.1 Fragmentation of Quarks

Even at the highest PETRA energies achieved so far the angular width of jets caused by the finite transverse momentum of roughly 350 MeV is still of the order of $2\delta \simeq 40°$. In order to infer the momentum of the primordial quark from experimental hadron distributions, or the other way around, to compare the predictions of perturbative QCD with measured hadron cross sections one needs phenomenological models which simulate this jet broadening, i.e. describe the fragmentation of quarks and gluons into hadrons in an as realistic way as possible. The hadronization effects are supposed to fall off with increasing energy W like 1/W, but with a coefficient that implies a rather important modification of perturbation theory predictions at presently accessible energies.

In this section we want to describe briefly the fragmentation of quarks. We encountered this problem already in connection with the interpretation of the data of the SLAC-LBL experiment in Sect.2.2 /Hanson et al., 1975,1982/. The authors of this paper described the quark jets in terms of a limited-transverse momentum jet model /van Hove, 1969/. In this model the phase space was modified by a matrix element squared of the form

$$M^2 = \exp\left[-\left(\sum_i \mathbf{p}_{iT}^2\right)/2b^2\right] \tag{2.4.1}$$

where $\mathbf{p}_{iT}$ is the momentum perpendicular to the jet axis for the ith particle and b is a parameter chosen to reproduce the average transverse momentum of 350 MeV observed in the data. The sum is over all produced particles. The assumed jet axis angular distribution was of the form (2.2.6). The generated events had only pions with the charged and neutral multiplicities given by separate Poisson distributions. Similar models have been worked out and applied to the interpretation of e^+e^- high energy data by Satz /1975/, Burrows /1979/, Eichmann and Steiner /1979/, Engels, Dabkowski and Schilling /1980/ and Clegg and Donnachie /1982/.

The model which is the most frequently used in data analysis, however, is the fragmentation model of Field and Feynman /1978/ which we shall describe in some-what more detail. This model is more flexible than the modified phase space model described above. In particular different quark flavours, all sorts of particles and the fragmentation of gluons can be incorporated quite easily. Furthermore its phe-nomenological construction rests much more on intuition based on quantum chromo-dynamics than the uncorrelated jet model.

Quarks and antiquarks in a hadron (meson = $q\bar{q}$, baryon = qqq) are usually consider-ed as relatively free to move around in a region of the size of a hadron of 1 fm $\simeq$ $(0.2 \text{ GeV})^{-1}$. In the process $e^+e^- \rightarrow q\bar{q}$, where q stands for u, d, s, c, b quarks, the quark and the antiquarks aquire a large momentum $\pm W/2$, sufficient to escape from the production point. This would result in the appearance of fractional charged par-ticles, which, however are not observed. This is usually explained by stating that the quarks are confined in the hadron, i.e. the forces $[V(r) \sim r$ or $F \sim const.]$ pre-vent the escape of the quarks and are responsible for sorting out quantum numbers so that only colour-singlet objects appear in the final state. Because of confine-ment, which is certainly a non-perturbative phenomenon, a string of gluon fields develops between quark and antiquark, which at later times breaks up into further quark-antiquark pairs, which then transform into mesons and baryons and their anti-particles. The final result is the transformation (or fragmentation) of the quark into a jet of hadrons. Thus the appearance of hadron jets (as opposed to free quarks) and their properties is a consequence of quark confinement and cannot be described by perturbative QCD. In the framework of QCD perturbation theory we can predict on-ly the distribution of quarks, antiquarks and gluons in the primordial production process at small distances and times and can say nothing about the transformation of these quark, antiquarks and gluons into hadrons observed as final states in e^+e^- annihilation.

In the Field-Feynman model the jet formation proceeds through a succession of fundamental breakups of the form

$$\text{quark} \rightarrow \text{meson} + \text{quark} \tag{2.4.2}$$

in which the meson containing the original quark and a new quark jet is formed which breaks up in the same way. The first quark q_0 with momentum p_0 is combined with an antiquark $\bar{q}_1$, which together with a quark q_1 is produced out of the vacuum (equiva-lent to breakup of the string), to a meson (= $q_0\bar{q}_1$) with momentum P_0 leaving the

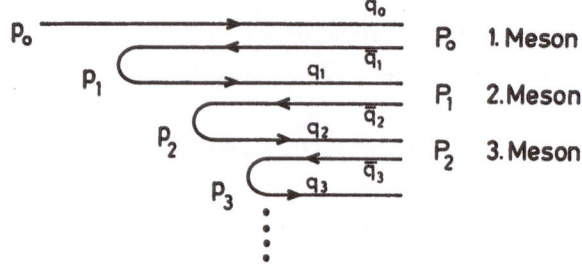

Fig.2.15. Hadronization cascade of a quark in the model of Field and Feynman

quark q_1 with momentum $p_1 = p_0 - P_0$ behind, which then breaks up in an equivalent fashion. This is shown schematically in Fig.2.15. At each step the quark momentum decreases. When it falls below a critical value $\simeq m_\pi$ the process stops and the last quark has no momentum left to escape. Then the jet is complete.

One assumes that the distribution of the relative momenta $z = P_i / p_i$ is independent at which point of the chain the fragmentation occurs and is described completely phenomenologically by the ansatz

$$f(z) = 1 - a + 3a(1 - z)^2 \quad . \tag{2.4.3}$$

Actually the variable used in practical calculations is not P/p but

$$z = \frac{(E + p_{||})_{meson}}{(E + p_{||})_{quark}} \quad . \tag{2.4.4}$$

This way rapidity distributions at different jet energies can be compared more easily. The transverse momenta p_{iT} of the quarks in the chain are calculated from the distribution

$$d\sigma/dp_T^2 \sim \exp(-p_T^2/2\sigma_q^2) \tag{2.4.5}$$

in such a way that the transverse momenta of q_i and $\bar{q}_i$ compensate each other. The transverse momentum p_{iT} of the mesons follows from the sum of the transverse momenta of the built in quarks and antiquarks and has the average value $<P_{iT}^2> = 2\sigma_q^2$.

Further free parameters in the Field-Feynman model are the ratio for the spins of the $q\bar{q}$ pair in the meson, which is

$$r = \frac{pseudoscalar\ mesons}{pseudoscalar + vector\ mesons} \tag{2.4.6}$$

(resonances with higher spin than $S = 1$ have not been taken into account yet) and the ratio of flavours for the sea quarks q_1, q_2, ... which are chosen according to $u : d : s : c : b = 2 : 2 : 1 : 0 : 0$. The parameters a, σ_q and r were fixed in the original work of Field and Feynman by fitting inelastic lepton-nucleon scattering data. For applications to e^+e^- annihilation these parameters are fitted again. The resulting values differ somewhat from the original Field-Feynman parameters and are: a = 0.5, σ_q = 0.33 GeV, r = 0.5. It is important that these parameters are determined from lower energy data (W < 22 GeV) or if higher energy data are used, the fit is performed only in such regions of phase space where 2-jet production dominates.

Recently the fragmentation of heavy quarks, c and b quarks, has been investigated in some detail from the theoretical /Bowler, 1981; Peterson, Schlatter, Schmitt and Zerwas, 1983; Jones and Migneron, 1983; Anderson, Gustafson and Söderberg, 1983/ and the experimental side /Althoff et al., 1983a; Adeva et al., 1983a/.

2.4.2 Phenomenological Models for Quark and Gluon Fragmentation

In the parton model, which is identical to zeroth order QCD, e^+e^- annihilation into hadron starts with the production of a quark q_0 and an antiquark $\bar{q}_0$ with a specific flavour q_0 = u, d, s, c and b and equal and opposite directed momentum. Which flavours are excited depends on the center-of-mass energies. Above W = 11 GeV all five flavours can be produced (see Fig.2.2). In the Field-Feynman model the quark q_0 and antiquark $\bar{q}_0$ fragment independently of each other in the way as described in Sect. 2.4.1. The cascade ends, when the remaining energy does not suffice for producing one more pion or kaon. The colour charge and other quantum numbers of the last quark are compensated with the last antiquark in the q_0 cascade which then gives the last meson produced, so that the picture looks like Fig.2.16.

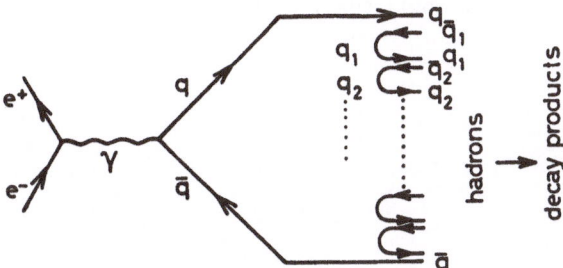

Fig.2.16. $e^+e^- \rightarrow q\bar{q}$ with subsequent fragmentation into stable and decaying hadrons

In the production $e^+e^- \rightarrow c\bar{c}$ and $e^+e^- \rightarrow b\bar{b}$ the finite quark masses are taken into account. They fragment in D and D* and B and B* mesons respectively. The masses of the B and B* mesons were calculated from the quark masses. Some of these masses have been measured recently /Behrends et al., 1983/, so that the theoretical values can be revised. The lowest D and B mesons decay weakly. The decays are also built into the Monte Carlo programs, either on the basis of empirical information /Particle

Data Group, 1982/ or using model calculations based on the weak decay of quasi-free quarks, in particular for the weak decays of b quarks $b \rightarrow c + \bar{u} + d$, $c + \bar{c} + s$ etc. including semileptonic decays /Ali, 1979; Ali, Körner, Kramer and Willrodt, 1979a,b, c,d;1980/. In more recent applications the fragmentation of q_0 and $\bar{q}_0$ into baryons is also taken into account on the basis of simple model considerations /Meyer, 1982/. All this input is built into the Monte Carlo program and complete final states giving all the hadrons and photons with their momenta and quantum numbers are produced. These are compared with experimental data and from this the free parameters like a, σ_q, r etc. of the fragmentation model are fixed /Brandelik et al., 1980a/. With the complete event structure at ones disposal one can also calculate the distributions in jet measures like thrust T, sphericity S, spherocity S', acoplanarity A etc. which were defined in Sect.2.3. As an example we present in Fig.2.17 thrust distributions for various center-of-mass energies W calculated with the Field-Feynman model for u, d, s quark production only. Therefore these curves do not include effects from weak decays of c and b quarks /Hoyer, Osland, Sander, Walsh and Zerwas, 1979; Kramer, 1980/. These T distributions vanish for $T \rightarrow 1$ caused by the non-vanishing masses of the produced hadrons. Furthermore they become narrower and narrower with increasing W as one expects since

$$<1-T> \simeq \frac{<p_T> <n>}{2W} \qquad (2.4.7)$$

so that the width of the curves should go down like <n>/W. We should notice, however, that for W = 10 GeV the width of the T distribution is quite large which means that at this low energy one cannot detect a third jet, like the gluon jet. This leads us already to the topic of the next chapter where we shall consider the production of 3 jets: $e^+e^- \rightarrow q\bar{q}g$ in detail.

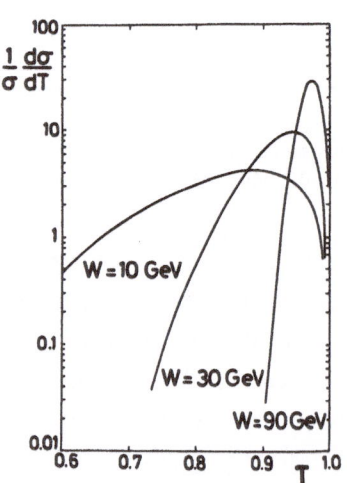

Fig.2.17. Thrust distribution $(1/\sigma)d\sigma/dT$ for c.m. energies W = 10, 30 and 90 GeV based on $e^+e^- \rightarrow q\bar{q}$ and Field-Feynman fragmentation without weak decays /Kramer, 1980/

In order to establish the existence of a third jet, the gluon jet, one needs a model for the hadronization of the gluon. For this purpose three different models have been developed, the model of Hoyer, Osland, Sander, Walsh and Zerwas /1979/, the model of Ali, Pietarinen, Kramer and Willrodt /1980/ and the model of Andersson, Gustafson and Sjöstrand /1980/. These three models, which have been used quite extensively for the interpretation of high energy e^+e^- hadron production will be designated as Hoyer, Ali and Lund model. In the following we shall describe these models in some detail in order to understand differences in the interpretation of experimental data.

The basic formulas, from which the contribution of the quark, antiquark and gluon intermediate state ($q\bar{q}g$) to the hadron final state is calculated, will be considered in the next chapter. There we shall show, that the $q\bar{q}g$ formula cannot be applied to the whole phase space because it contains divergences due to infrared and mass singularities. Therefore one must introduce parameters which have the purpose to define the multiplicities for 2 and 3 jets in the order g^2 of the quark-gluon coupling. In the Hoyer and in the Ali model this is achieved by introducing a thrust cut-off $T_0 = 0.95$. This means that only for thrust values $T \leq T_0$, where T is the thrust of the $q\bar{q}g$ state, the q, $\bar{q}$ and g fragment independently of each other in a way as it was described in Sect.2.4.1 for the quark alone. For the remainder of the $q\bar{q}g$ phase space with $T_0 < T \leq 1$ it is assumed that it is part of the 2-jet contribution which is calculated from $e^+e^- \to q\bar{q}$ with Field-Feynman fragmentation as described in the previous section. Then depending on the quark-gluon coupling g the 3-jet multiplicity is of the order of 30%. In some applications the $q\bar{q}g$ contribution is reduced further by demanding that each parton has at least an energy of 2 GeV (see also Odorico /1980/).

In particular, in the Hoyer model some further simplifying assumptions are made for the fragmentation of the $q\bar{q}g$ state. The $q\bar{q}$ part in the $q\bar{q}g$ state fragments in the same way as the $q\bar{q}$ state in $e^+e^- \to q\bar{q}$ according to the Field-Feynman model except that the momenta of q and $\bar{q}$ are different and are determined by the cross section for $e^+e^- \to q\bar{q}g$. Furthermore it is assumed that the gluon fragments independently like a single quark with flavour neutrality achieved by selecting the right mixture of quark flavours. The fragmentation parameter a in (2.4.3) for the gluon is $a(g) = 1$ which makes the gluon fragmentation softer than that of the quark as one expects it.

The model of Ali et al. differs from the Hoyer model mostly in the gluon fragmentation. Here the gluon splits into a $q\bar{q}$ pair $g \to q\bar{q}$ with the splitting function

$$f_g(z) = \frac{1}{2}[z^2+(1-z)^2]$$ (2.4.8)

which is motivated by perturbative QCD (Altarelli-Parisi splitting function /Altarelli and Parisi, 1977/). Then the $q\bar{q}$ system produced by the gluon fragments according to the Field-Feynman model with σ_g chosen as 0.33 GeV. Recent experiments of the JADE Collaboration at PETRA /Bartel et al., 1983/ indicate, however, that the average transverse momentum of hadrons originating from the gluon jet is somewhat larger than those coming from quark jets. In addition the scaling violation of the fragmentation function (2.4.3) as predicted by QCD in the leading logarithm approximation is built in. But this turned out not very important. Furthermore the contribution of 4 parton production: $e^+e^- \to q\bar{q}gg$ and $e^+e^- \to q\bar{q}q\bar{q}$ was taken into account. These contributions were allowed to fragment in the same fashion as the $q\bar{q}g$ state. Similarly to the $q\bar{q}g$ state only those parts of the $q\bar{q}gg$ and $q\bar{q}q\bar{q}$ intermediate state were included which produce genuine 4-jet final states. This was achieved by introducing a cut-off in the acoplanarity A, which was A_0 = 0.05. This way the infrared and mass singular contributions degenerate with 2 and 3 jets are excluded.

In the model of Andersson, Gustafson and Sjöstrand the gluon fragmentation is based on a consequent application of the string picture. Between the produced quark and antiquark the field lines of the gluon field are squeezed into a string. Mesons and baryons with their masses and transverse momenta are produced by breaking up this string. The gluon in the $q\bar{q}g$ intermediate state is considered as a local transveral excitation of the string, which is connected with q and $\bar{q}$ via gluon field lines. This local excitation breaks up by emission of a fast meson in an antiquark $\bar{q}''$ and a quark q' which take over the rest of the gluon momentum. $\bar{q}''$ and q' are connected through gluon field lines with the original quark q and antiquark $\bar{q}$ so that two new strings $q\bar{q}''$ and $q'\bar{q}$ are formed. They break up into mesons (or baryons if considered) in their respective center-of-mass systems. The fragmentation, in particular the limitation of the transverse momentum, follows the field lines between $q\bar{q}''$ and $\bar{q}q'$ but not directly between q and $\bar{q}$. This way the produced particles have their momenta lie along hyperbolas between the three jets, qg, $g\bar{q}$ but not $q\bar{q}$, as is shown in Fig.2.18. After the fragmentation the momenta of the produced particles are transformed back from the $q\bar{q}''$ and $\bar{q}q'$ center-of-mass systems to the e^+e^- center-of-mass system. This choice of a different coordinate system for the fragmentation has the effect that more particles are emitted in the direction of the quark and antiquark momentum and less in

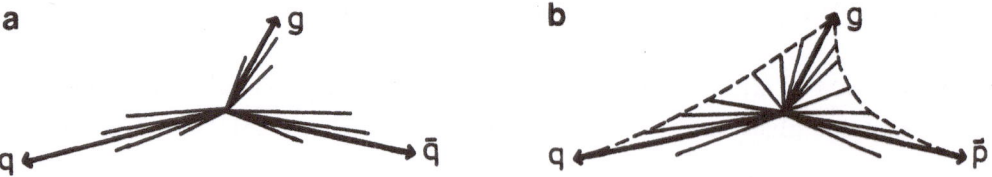

Fig.2.18. a) Independent fragmentation of $q\bar{q}g$ as in the Hoyer and Ali model and b) fragmentation along strings as in the Lund model

the direction of the gluon momentum compared to the other two models /Hoyer et al., 1979, and Ali et al., 1980/ where quarks and gluons fragment completely independently of each other (see also Montvay /1979/ who considers a completely arbitrary coordinate system). This effect increases with the masses of the produced hadrons. For massless particles, i.e. practically also pions, the Lorentz transformations between the two center-of-mass systems have less effect than for particles with larger masses like kaons and protons.

The limitation of the 3-jet multiplicity is achieved in the Lund model somewhat differently than in the Hoyer and Ali model. The parameters for the calculation of the 3-jet rate are based on the string picture and are such that the 3-jet rate lies near 60%. Furthermore in the original work the longitudinal fragmentation is not given by (2.4.3) but by other more complicated functions which are justified by one-dimensional QCD and the string picture /Andersson, Gustafson and Peterson, 1979; Andersson and Gustafson, 1980; Andersson, Gustafson and Sjöstrand, 1982/. In most of the analyses with the Lund model done by experimentalists these functions are again replaced by (2.4.3).

In all three fragmentation models one basic ingredient is the impulse approximation. This means that the *probabilities* for the emission of mesons and baryons calculated on the basis of given distributions like (2.4.3) and (2.4.5) are added together and not the amplitudes for the production of particles from primordial quarks and gluons as the laws of quantum theory demand it. Presumably the effect of neglecting interference terms is small for the production of a large number of particles. But this has never been checked. The summation of probabilities can be justified for the emission of soft gluons in the leading logarithm approximation of perturbative QCD. Whether this approximation is relevant for the hadronization of quarks and gluons is still an open question. Several authors have developed hadronization models on that basis: Odorico /1980/; Mazzanti and Odorico /1980/; Fox and Wolfram /1980/; Ritter /1982/; Field and Wolfram /1982/; Gottschalk /1982/. They can be considered as an alternative to the more phenomenological models described above. Which is the right approach can be found out only by detailed comparisons with experimental data. This has been done so far only for the phenomenological models. For the QCD motivated models which are somewhat less flexible this still has to be done.

3. e⁺e⁻ Annihilation into Jets in QCD Perturbation Theory

3.1 Jets in Order α_s

3.1.1 Introduction

In this chapter we shall consider all Feynman diagrams which contribute to cross sections for the annihilation of e^+e^- into quarks and gluons, i.e. $e^+e^- \rightarrow q\bar{q}$, $q\bar{q}g$, $q\bar{q}gg$ and $q\bar{q}q\bar{q}$. We shall explain how the contributions of all these diagrams should be interpreted, in particular, to which jet multiplicity they belong and how the cross sections for a specific number of jets look like.

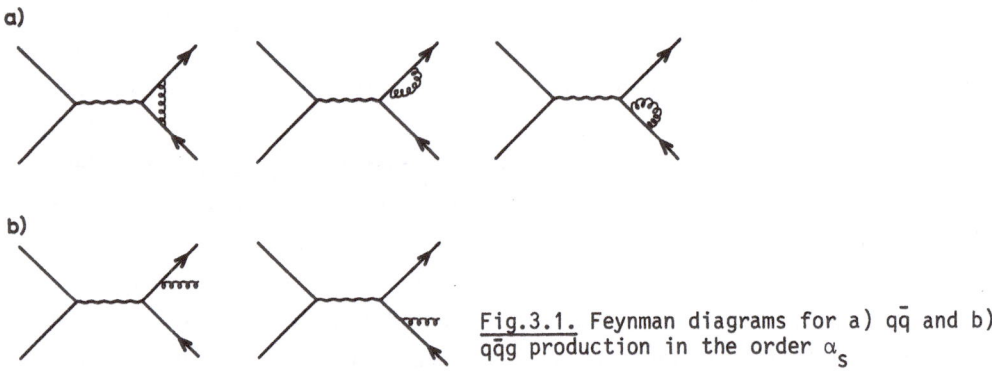

Fig.3.1. Feynman diagrams for a) $q\bar{q}$ and b) $q\bar{q}g$ production in the order α_s

   We start with the diagrams which contribute in the order g^2 where g is the quark-gluon-coupling constant and $\alpha_s = g^2/4\pi$. These diagrams are represented in Fig.3.1a,b. The graphs in Fig.3.1a represent the virtual corrections to the amplitude $e^+e^- \rightarrow q\bar{q}$. They are $O(\alpha_s)$ and are multiplied with the amplitude $e^+e^- \rightarrow q\bar{q}$ in zeroth order of g (see Fig.2.1) so that a contribution to the cross section for $e^+e^- \rightarrow qq$ in $O(\alpha_s)$ results. The diagrams in Fig.3.1a, therefore, determine the $O(\alpha_s)$ corrections to the 2-jet cross section $e^+e^- \rightarrow q\bar{q}$. As is well known each of the diagrams in Fig.3.1a is ultraviolet divergent, i.e. the integrals over the virtual momenta k diverge for $k \rightarrow \infty$. This divergence cancels, however, if the contributions of all three diagrams are added. This must be so, since the electromagnetic charge of the quarks is not renormalized by the quark-gluon interaction. The three diagrams remain divergent nevertheless because of the behaviour of their integrands for $k \to 0$, which is the well-known infrared divergence caused by the masslessness of the gluons. Such infrared divergences appear in perturbation theory for all theories with massless vector particles. These infrared divergences cancel if one adds those contributions from the process $e^+e^- \to q\bar{q}g$ which have also these infrared divergences. This will be discussed later. Because we shall assume that also the quarks are massless, which is certainly justified for u, d, s quarks, a further divergence appears, the so-called mass divergence or collinear divergence. This singularity comes about because the virtual gluon (with momentum k) in Fig.3.1a can be collinear with the outgoing quark (with momentum p_1) so that the invariant scalar product

$$2kp_1 = 2|\mathbf{k}|\,|\mathbf{p}_1|\,(1-\cos\theta) \tag{3.1.1}$$

which appears in the denominator of the quark propagator vanishes for $\theta \to 0$.

To control these divergences a regularization procedure to define the divergent integrals must be adopted. The most convenient method, which has the advantage to maintain the gauge invariance, is the dimensional regularization procedure of 't Hooft and Veltman /1972/ (see also Bollini and Giambiagi /1971/). In this method the Feynman diagrams are calculated for arbitrary dimension n (with n > 4). Then the infrared and mass singularities appear as poles in 4-n. Useful references for this method are Mariciano /1975/ and Leibrandt /1975/.

Other graphs which must be considered for $e^+e^- \to$ quarks + gluons up to order α_s are the well-known gluon-bremsstrahlung diagrams in Fig.3.1b. From these the cross section for $e^+e^- \to q\bar{q}g$ in lowest order α_s is calculated. $q\bar{q}g$ is now a final state consisting of 3 jets, a quark, an antiquark and a gluon jet. The cross section for this process was calculated the first time by Ellis, Gaillard and Ross /1976,1977/. In the following section we shall give the results for various cross sections for $e^+e^- \to q\bar{q}g$ and shall derive several distributions of interest from it.

3.1.2 Cross Section for $e^+e^- \to q\bar{q}g$

The kinematics for $e^+e^- \to q\bar{q}g$ is the same as the kinematics for the decay of a virtual photon of mass W into three massless particles. Let us denote the four vectors of q, $\bar{q}$ and g by p_1, p_2 and p_3, respectively. The virtual photon has momentum q, so that $q = p_1 + p_2 + p_3$, in particular, in the e^+e^- center-of-mass system $\mathbf{p}_1 + \mathbf{p}_2 + \mathbf{p}_3 = 0$, $p_{10} + p_{20} + p_{30} = q_0 = W(q^2 = W^2)$ and $|\mathbf{p}_i| = p_{i0}$ for massless quarks. Instead of p_{i0} we use the dimensionless variable

$$x_i = 2p_{i0}/W, \quad i = 1, 2, 3 \quad . \tag{3.1.2}$$

Then $x_1 + x_2 + x_3 = 2$ and $0 \leq x_i \leq 1$. Only two of the x_i are independent. Let us take x_1 and x_2. The kinematically allowed region is the triangle $0 \leq x_1 \leq 1$ and $1 - x_1 \leq x_2 \leq 1$ in Fig.3.2. All $q\bar{q}g$ events must be in this triangle. It defines the Dalitz plot for the decay of the virtual photon with mass W into the three massless quanta q, $\bar{q}$ and g, $\gamma \to q\bar{q}g$. The line $x_2 = 1 - x_1$ belongs to $x_3 = 1$.

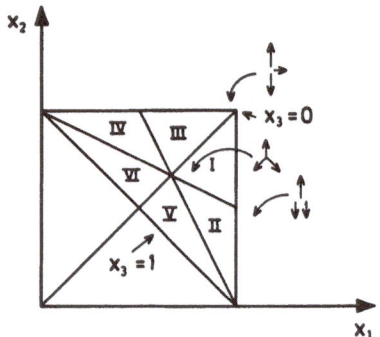

Fig.3.2. Phase space diagram for $e^+e^- \to q\bar{q}g$ with regions ordered according to magnitude of x_1, x_2 and x_3

Since all three jets, q, $\bar{q}$ and g, fragment into hadrons and only these hadrons are detected experimentally, it is, without further theoretical input, not possible to identify the jets as quark, antiquark or gluon jet. This means that for a genuine 3-jet event only the three jet energies x_1, x_2 and x_3 or the angles between the three jet momenta can be measured. In this case it is useful to order the three jets in such a way that always $x_1 > x_1 > x_3$. Of course, x_1 does not denote the energy of the quark jet anymore. It is the jet with the largest energy. With this ordering x_1 and x_2 vary only over the small triangle which is marked I in Fig.3.2. The other five triangles, II, III, IV, V and VI correspond to the other orderings $x_1 > x_3 > x_2$, $x_2 > x_1 > x_3$ etc., which need not to be considered anymore. In the triangle I we have near $x_1 \simeq 1$ those events where the 3 jets degenerate to 2 jets. In the upper corner, i.e. for $x_1 \simeq x_2 \simeq 1$ ($x_3 \simeq 0$) jet 1 and jet 2 are emitted with roughly equal and opposite momentum and jet 3 comes out only with a small momentum. At the point $x_1 \simeq 1$ and $x_2 \simeq 1/2$, so that also $x_3 \simeq 1/2$, jet 2 and jet 3 are produced collinear opposite in direction to jet 1. Only near the point with $x_1 \simeq x_2 \simeq x_3 \simeq 2/3$ lie the real interesting, genuine 3-jet events. The cross section for the production of these events is, however, much smaller as it will be apparent below than for the degenerate events near $x_1 \simeq 1$.

In the e^+e^- center-of-mass system, i.e. in the laboratory system of the storage ring, $\mathbf{q} = 0$, the momenta of the three jets q, $\bar{q}$ and g lie in a plane. For the complete description of the three-body final state one needs still two angles θ and χ

which give the orientation of the jet plane with respect to the e^+ (or e^-) beam direction.

The calculation of the cross section for

$$\gamma(q) \rightarrow q(p_1) + \bar{q}(p_2) + g(p_3) \tag{3.1.3}$$

starts from the Feynman diagrams in Fig.3.1b. The sum of the two diagrams yields the amplitude T_μ, where μ is the polarization of the virtual photon. From this one obtains the hadronic tensor

$$H_{\mu\nu} = \sum_{\text{spins, colours}} T_\mu T_\nu^* \quad . \tag{3.1.4}$$

The sum in (3.1.4) is over all spin and colour states of q, $\bar{q}$ and g which are unobserved. The result of the calculation is /Ellis, Gaillard and Ross, 1976/:

$$H_{\mu\nu} = 4Q_f^2 g^2 N_c C_F \left\{ \frac{1}{p_1 p_3} \left[\{p_2,p_3\}_{\mu\nu} - \{p_1,p_1\}_{\mu\nu} + \{p_1,p_2\}_{\mu\nu} \right] \right.$$

$$+ \frac{1}{p_2 p_3} \left[\{p_1,p_3\}_{\mu\nu} - \{p_2,p_2\}_{\mu\nu} + \{p_1,p_2\}_{\mu\nu} \right]$$

$$\left. + \frac{p_1 p_2}{p_1 p_3 p_2 p_3} \left[2\{p_1,p_2\}_{\mu\nu} + \{p_1,p_3\}_{\mu\nu} + \{p_2,p_3\}_{\mu\nu} \right] \right\} \tag{3.1.5}$$

where

$$\{p_i,p_j\}_{\mu\nu} = p_{i\mu}p_{j\nu} + p_{j\mu}p_{i\nu} - p_1 p_2 g_{\mu\nu} \quad . \tag{3.1.6}$$

Q_f is the quark charge with flavour f, $N_c = 3$ the number of colours and C_F is the Casimir operator which appears because of the λ matrices in the quark-gluon coupling. It is

$$C_F = \sum_a \frac{\lambda_a}{2} \frac{\lambda_a}{2} = \frac{N_c^2 - 1}{2N_c} = \frac{4}{3} \quad . \tag{3.1.7}$$

In (3.1.5) the quark masses $m_f = 0$. From (3.1.5) all cross sections of interest can be calculated. For this $H_{\mu\nu}$ is contracted with the well-known lepton tensor $L^{\mu\nu}$ for unpolarized electrons and positrons

$$L^{\mu\nu} = \{p_+,p_-\}^{\mu\nu} \tag{3.1.8}$$

where p_+ and p_- are the momenta of the positron and electron, respectively. For unpolarized initial leptons the cross section depends in general only on two angles θ and χ. In addition it depends on the two scaled momenta x_1 and x_2. The θ and χ dependence is well-known and has the following form /Hirshfeld and Kramer, 1974; Avram and Schiller, 1974/

$$2\pi \frac{d^4\sigma}{d\cos\theta d\chi dx_1 dx_2} = \frac{3}{8}(1+\cos^2\theta)\frac{d^2\sigma_U}{dx_1 dx_2} + \frac{3}{4}\sin^2\theta \frac{d^2\sigma_L}{dx_1 dx_2}$$

$$+ \frac{3}{4}\sin^2\theta\cos 2\chi \frac{d^2\sigma_T}{dx_1 dx_2} - \frac{3}{2\sqrt{2}}\sin 2\theta\cos\chi \frac{d^2\sigma_I}{dx_1 dx_2} \quad . \tag{3.1.9}$$

In (3.1.9) θ is the angle between the incoming electron beam and a vector determined by the final state, which will be specified below. χ is the azimuthal angle around this vector between the event plane and the plane determined by the beam axis and this specified vector of the final state. As vectors, which specify the event plane we can choose, for example, the momenta $\mathbf{p}_1$ and $\mathbf{p}_2$. Then either $\mathbf{p}_1$ is along the z axis and $\mathbf{p}_1 \times \mathbf{p}_2$ along the y axis, this defines the helicity system and θ and χ the helicity angles, or we can choose $\mathbf{p}_1 \times \mathbf{p}_2$ as z axis and p_1 parallel to the x axis (transversality system) /Avram and Schiller, 1974; Hirshfeld, Kramer and Schiller, 1974/. In the following we shall consider only the helicity system.

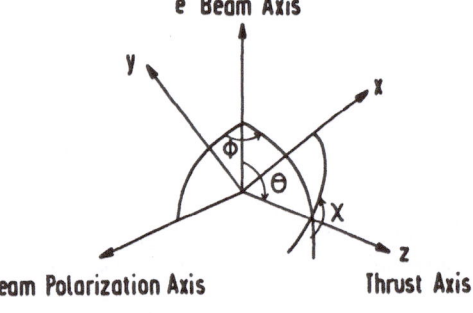

Fig.3.3. Definition of Euler angles θ, χ and ϕ. The event lies in the x,z plane. The thrust axis is along $\vec{Oz}$. Ox points in the direction of the second most energetic jet. $0 \leq \theta \leq \pi$, $0 \leq \chi \leq 2\pi$ and $0 \leq \phi \leq 2\pi$

The cross sections σ_U, σ_L, σ_T and σ_I have the following interpretation. $\sigma_U(\sigma_L)$ is the cross section for unpolarized transverse (longitudinally polarized) virtual photons with helicity axis $\vec{Oz}$ (see Fig.3.3, where z is along the thrust axis). $\sigma_T(\sigma_I)$ corresponds to the interference of helicity +1 and helicity -1 amplitudes (the real part of helicity +1 and 0 interference). When defining the coordinate system of the final state one must be aware of the fact that the momenta of the quanta of the final state q, $\bar{q}$ and g are not observed. Only the direction of one of the three hadron jets can be measured. Therefore the z axis of the helicity system is usually identified with the thrust axis. The axis $\vec{Oz}$ is chosen to point into the hemisphere in which one finds the second most energetic jet originating either from a quark, antiquark or gluon. In order to obtain the partial cross section $d\sigma_X$ (X = U, L, T and I) for such a special coordinate system we need the partial cross sections for the three cases: (i) $\mathbf{p_1}||\overrightarrow{0z}$, (ii) $\mathbf{p_2}||\overrightarrow{0z}$ and (iii) $\mathbf{p_3}||\overrightarrow{0z}$. The T was defined in (2.3.4). For 3 massless particles it is equal to the maximal momentum, i.e.

$$T = \max(x_1, x_2, x_3) \qquad\qquad (3.1.10)$$

and the thrust axis is the direction of the largest momentum. Therefore if we choose as variables T_1, T_2 and T_3, where the T_i denote the ordered x_i: $T_1 > T_2 > T_3$, we need the cross sections in three kinematical regions (see Fig.3.2): I + II : $x_1 > x_2, x_3$; III + IV : $x_2 > x_1, x_3$ and V + VI : $x_3 > x_1, x_2$. In region I + II (III + IV) the thrust axis coincides with the direction of the outgoing quark (antiquark) while in region V + VI the thrust axis corresponds to the gluon momentum.

Of course, the sum of σ_U and σ_L which we denote by $\sigma = \sigma_U + \sigma_L$ must be independent of the choice of the coordinate system. It is /Ellis, Gaillard and Ross, 1976, 1977/

$$\frac{d^2\sigma}{dx_1 dx_2} = \sigma^{(2)} \frac{\alpha_s}{2\pi} C_F \frac{x_1^2 + x_2^2}{(1-x_1)(1-x_2)} \qquad\qquad (3.1.11)$$

where $\sigma^{(2)}$ is the $q\bar{q}$ cross section in zeroth order from Sect.2.1:

$$\sigma^{(2)} = \frac{4\pi\alpha^2}{3q^2} N_C \sum_f Q_f^2 \quad . \qquad\qquad (3.1.12)$$

The other partial cross sections are /Kramer, Schierholz and Willrodt, 1978,1979/:
(i) $\mathbf{p_1}||\overrightarrow{0z}$:

$$\frac{d^2\sigma_L}{dx_1 dx_2} = \sigma^{(2)} \frac{\alpha_s}{2\pi} C_F \frac{2(1-x_3)}{x_1^2}$$

$$\frac{d^2\sigma_T}{dx_1 dx_2} = \frac{1}{2} \frac{d^2\sigma_L}{dx_1 dx_2}$$

$$\frac{d^2\sigma_I}{dx_1 dx_2} = \sigma^{(2)} \frac{\alpha_s}{2\pi} C_F \left(\frac{1-x_3}{2(1-x_1)(1-x_2)}\right)^{1/2} \frac{x_1 x_2 - 2(1-x_3)}{x_1^2} \qquad\qquad (3.1.13)$$

(ii) $\mathbf{p_2}||\overrightarrow{0z}$ = (i) with $x_1 \leftrightarrow x_2$ $\qquad\qquad (3.1.14)$

(iii) $\mathbf{p}_3 || \vec{Oz}$

$$\frac{d^2\sigma_L}{dx_1 dx_2} = \sigma^{(2)} \frac{\alpha_s}{2\pi} C_F \frac{4(1-x_3)}{x_3^2}$$

$$\frac{d^2\sigma_T}{dx_1 dx_2} = \frac{1}{2} \frac{d^2\sigma_L}{dx_1 dx_2}$$

$$\frac{d^2\sigma_I}{dx_1 dx_2} = \sigma^{(2)} \frac{\alpha_s}{2\pi} \left(\frac{1-x_3}{2(1-x_1)(1-x_2)} \right)^{1/2} \frac{x_1^2 - x_2^2}{x_3^2} \quad . \tag{3.1.15}$$

In these formulae the x_i ($i = 1, 2, 3$) are the scaled momenta for quark, antiquark and gluon. Only the vectors which define the hadron production plane have been chosen differently, for (i) $\mathbf{p}_1$ is along the z axis, $\mathbf{p}_1 \times \mathbf{p}_2$ along the y axis, for (ii) $\mathbf{p}_1$ and $\mathbf{p}_2$ are interchanged and for (iii) $\mathbf{p}_3$ is along the z axis and $\mathbf{p}_3 \times \mathbf{p}_1$ along the y axis. Note that it is the term proportional to $\cos\chi$ in the angular distribution (3.1.9) which, being asymmetric in $\chi \to \chi + \pi$ (i.e. $\vec{Ox} \to -\vec{Ox}$), makes a proper definition of χ necessary.

It is a simple matter to derive from (3.1.11-15) the cross sections as a function of the jet variables T_1, T_2 and T_3, where T_1 is the scaled energy of the most energetic jet [= thrust according to (3.1.10)], T_2 the scaled energy of the second most energetic jet and T_3 of the least energetic jet. For example, for $d\sigma = d\sigma_U + d\sigma_L$ this distribution in jet variables can be read off from (3.1.11) and is:

$$\frac{d^2}{dT_1 dT_2} = \sigma^{(2)} \frac{\alpha_s}{\pi} C_F 2 \left\{ \frac{T_1^2 + T_2^2}{(1-T_1)(1-T_2)} + \frac{T_2^2 + T_3^2}{(1-T_2)(1-T_3)} + \frac{T_3^2 + T_1^2}{(1-T_3)(1-T_1)} \right\} \quad . \tag{3.1.16}$$

Of course, T_1 and T_2 now vary only over the triangle I in Fig.3.2. Expressions for the other cross sections can be found in the paper by Schiller and Zech /1982/.

The formulae (3.1.11-16) are derived with quark masses $m = 0$. The equivalent results for $m \neq 0$ have been derived by Kramer, Schierholz and Willrodt /1980/ and Laermann and Zerwas /1980/. These formulae must be applied for $e^+e^- \to c\bar{c}g$ or $e^+e^- \to b\bar{b}g$ near threshold where quark masses cannot be neglected.

The most important formula for the interpretation of experimental data so far is (3.1.11). From this formula all distributions, which have been considered in connection with the interpretation of 3-jet events as quark-antiquark-gluon production, can be derived. Up to now, mostly only one-dimensional distributions have been considered, since the statistics of the data was not sufficient to test two-dimensional distributions. The other partial cross sections like $d\sigma_L$, $d\sigma_T$ and $d\sigma_I$ have not been measured yet. Some preliminary data for $d\sigma_L$ exist which will be considered below.

Before we go on we would like to point out that the cross section (3.1.11) has the most singular behaviour if x_1 and/or $x_2 \to 1$. The singularity for x_1 and $x_2 \to 1$ comes from the infrared singular behaviour ($x_3 \to 0$) whereas the behaviour for $x_1 \to 1$ or $x_2 \to 1$ contains the mass or collinear singularity. Further below we shall discuss how these singular contributions cancel against the singularities in the virtual diagrams.

In the following section we shall discuss some one-dimensional distributions which are derived from (3.1.11-15).

3.1.3 Thrust Distributions

As has been remarked already only cross sections for jet variables can be measured since quark-antiquark and gluon jets cannot be distinguished. Such a jet variable is the thrust T introduced in Sect.2.3 and which for the 3 particle final state is identical to $T_1 = \max(x_1, x_2, x_3)$. To obtain the distribution in $T \equiv T_1$ we must integrate over the second variable in the appropriate kinematical domain. The easiest case is $d\sigma/dT = d\sigma_U/dT + d\sigma_L/dT$ which can be deduced from (3.1.16) by integrating over T_2 in the triangle I of Fig.3.2. But we can start also with (3.1.11) and identify T in the various regions I to VI in Fig.3.2 and integrate over the remaining variable. In I and II we have $T = x_1$ and $2(1-T) \leq x_2 \leq T$, in III and IV x_1 and x_2 are interchanged, i.e. $T = x_2$ and $2(1-T) \leq x_1 \leq T$ whereas in V and VI we have $T = x_3$ and $2(1-T) \leq x_2 \leq T$. Therefore the thrust distribution is obtained from the integrals:

$$\frac{1}{\sigma^{(2)}} \frac{d\sigma}{dT} = \frac{\alpha_s}{2\pi} C_F \left\{ 2 \int_{2(1-T)}^{T} dx_2 \frac{T^2 + x_2^2}{(1-T)(1-x_2)} + \int_{2(1-T)}^{T} dx_2 \frac{(2-T-x_2)^2 + x_2^2}{(x_2 + T - 1)(1-x_2)} \right\} . \quad (3.1.17)$$

The result of the integration is /De Rujula, Ellis, Floratos and Gaillard, 1978/:

$$\frac{1}{\sigma^{(2)}} \frac{d\sigma}{dT} = \frac{\alpha_s}{2\pi} C_F \left\{ \frac{2(3T^2 - 3T + 2)}{T(1-T)} \ln \frac{2T-1}{1-T} - \frac{3(3T-2)(2-T)}{1-T} \right\} . \quad (3.1.18)$$

(3.1.18) is plotted in Fig.3.4. We see that $d\sigma/dT$ diverges for $T = 1$ and decreases strongly with decreasing T. At the kinematic boundary $T = 2/3$ (which is the boundary of I + II in Fig.3.2) $d\sigma/dT$ vanishes. The singular behaviour for $T \to 1$ is given by

$$\frac{1}{\sigma^{(2)}} \frac{d\sigma}{dT}_{T \to 1} = \frac{2\alpha_s}{\pi} C_F \frac{1}{1-T} \ln|1-T| . \quad (3.1.19)$$

The singularity at $T = 1$ has its origin in the singular behaviour of (3.1.11) which diverges for $x_1 \to 1$ and/or $x_2 \to 1$.

41

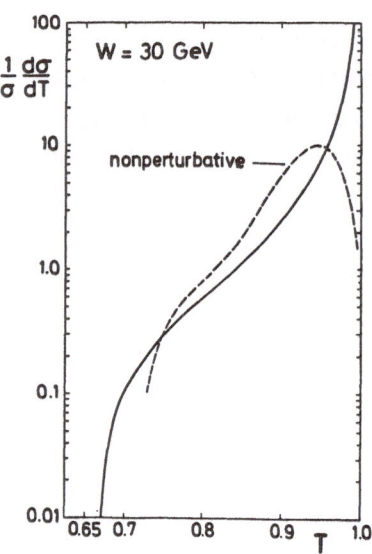

$\frac{1}{\sigma}\frac{d\sigma}{dT}$

W = 30 GeV

nonperturbative

Fig.3.4. Thrust distribution $(1/\sigma)d\sigma/dT$ at W = 30 GeV for $e^+e^- \to q\bar{q}g$ in the order α_S ($\alpha_S = 0.2$) compared to non-perturbative thrust distribution for 2-jet production based on $e^+e^- \to q\bar{q}$ (q = u, d, s, c, b) including broadening caused by weak decays of c and b quarks

Since (3.1.11) was derived in perturbation theory as the first term in an expansion in α_s the formula can be applied only for such regions of the kinematic variables x_1 and x_2 for which $d^2\sigma/dx_1dx_2$ is not too large. Otherwise the integrated 3-jet cross section could come out larger than the total cross section which is roughly equal to $\sigma^{(2)}$. From this consideration one obtains limits for the variables x_1 and x_2: x_1, $x_2 \leq x_0 < 1$ and similarly for T: $T \leq T_0 < 1$. These limits, i.e. x_0 and T_0, will be discussed later in connection with the 2-jet and integrated 3-jet cross section. It should be clear that it makes much more sense to apply (3.1.11) and (3.1.18) in such kinematic regions which are farther away from the infrared singular regions x_1, $x_2 = 1$ and T = 1, respectively. In this region, unfortunately the cross section is rather low (see Fig.3.4). The region x_1, $x_2 \leq x_0$ and $T \leq T_0$, respectively, is the genuine 3-jet region and the integral over this region is the integrated 3-jet cross section. Of course, this cross section depends on how the boundary is defined (either x_1, $x_2 \leq x_0$ or $T \leq T_0$ or in other ways).

Whereas $d\sigma/dT$ is not finite for T = 1 and also cannot be integrated up to T = 1 mean values of $(1-T)^n$, n = 1, 2, ..., are finite quantities. For example, the mean value of $1-T$

$$<1-T> = \frac{1}{\sigma^{(2)}} \int_{2/3}^{1} dT \frac{d\sigma}{dT} (1-T) \tag{3.1.20}$$

with $d\sigma/dT$ given by (3.1.18) is:

$$<1-T> = \frac{\alpha_s}{2\pi} C_F \left\{ -\frac{3}{4} \ln 3 - \frac{1}{18} + 4 \int_{2/3}^{1} dT \frac{1}{T} \ln\left(\frac{2T-1}{1-T}\right) \right\} = 1.05 \frac{\alpha_s}{\pi} \quad . \tag{3.1.21}$$

In <1-T> as defined by (3.1.20) the singularity $(1-T)^{-1}$ coming from $d\sigma/dT$ cancels. The remaining singularity for $T = 1$ is integrable and a finite value for <1-T> follows. Therefore one says <1-T> is an "infrared save" quantity in contrast to the integrals over $d\sigma/dT$ up to $T = 1$. In the same way all higher powers of 1-T have finite mean values. One should notice, however, that the mean value as defined in (3.1.20) is not the mean value in the usual sense, since (3.1.20) is not the mean value of (1-T) with the distribution (3.1.18). This would be equal to zero since the integral over (3.1.18) which usually appears in the denominator is infinite. Therefore $(1-T)d\sigma/dT$ in (3.1.20) was normalized with $\sigma^{(2)}$. The interpretation of such mean values as (3.1.20) will be discussed later after the results for the 2-jet cross section up to order α_s have been presented.

Next we consider the other partial cross sections $d\sigma_X$ (X = L, T and I) in (3.1.13-15) and integrate over one variable. To obtain the thrust distributions one has to superimpose the cross sections from (i), (ii) and (iii) in their appropriate regions in such a form that the z axis of the event plane is always the thrust axis (see Fig. 3.3). The result of this integration is /Kramer, Schierholz and Willrodt, 1978,1979/:

$$\frac{1}{\sigma^{(2)}} \frac{d\sigma_L}{dT} = \frac{\alpha_s}{2\pi} C_F \frac{2(8T-3T^2-4)}{T^2}$$

$$\frac{1}{\sigma^{(2)}} \frac{d\sigma_T}{dT} = \frac{1}{2} \frac{1}{\sigma^{(2)}} \frac{d\sigma_L}{dT} \qquad\qquad (3.1.22)$$

$$\frac{1}{\sigma^{(2)}} \frac{d\sigma_I}{dT} = \frac{\alpha_s}{2\pi} C_F \sqrt{2} \left(T^2-2T+2\right)\left(\frac{2}{T^2} \sqrt{2T-1} - \frac{1}{T\sqrt{1-T}}\right) \ .$$

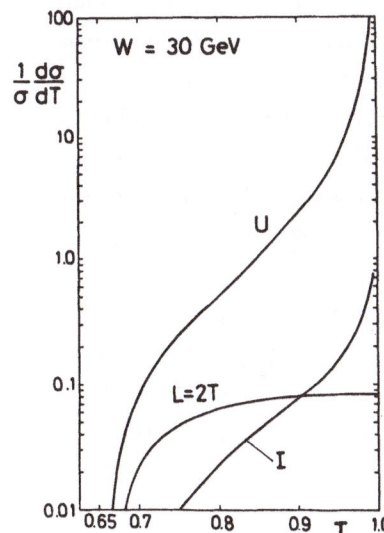

Fig.3.5. Cross sections dependent on polarization of virtual photon: U: $(1/\sigma)d\sigma_U/dT$, L: $(1/\sigma)d\sigma_L/dT$, I: $(1/\sigma)(-d\sigma_I/dT)$, $d\sigma_T/dT = (1/2)d\sigma_L/dT$; for W = 30 GeV and $\alpha_s = 0.2$

These cross sections have to be inserted into (3.1.9) if one wants the full cross section $d^3\sigma/d\cos\theta d\chi dT$. Of course, $d\sigma_U/dT = d\sigma/dT - d\sigma_L/dT$. The partial cross sections (3.1.22) are plotted in Fig.3.5. We see that only $d\sigma_U/dT$ has the original infrared singular behaviour for $T \to 1$. $d\sigma_L/dT$ is finite for $T = 1$ and in general much smaller than $d\sigma_U/dT$. $d\sigma_I/dT$ is singular at $T = 1$. But the singularity is integrable. In $O(\alpha_s)$ we have the relation

$$d\sigma_T/dT = \frac{1}{2}\, d\sigma_L/dT \qquad\qquad\qquad\qquad\qquad (3.1.23)$$

which follows from the same relation for the double differential cross sections in (3.1.13-15).

The measurement of the angular distribution (3.1.9) with the cross sections $d\sigma_I/dT$ (X = U, L, T and I) in (3.1.18) and (3.1.22) is a further independent way, to verify the existence of the $q\bar{q}g$ final state in jet production. This way more details of the hadron tensor (3.1.5) are tested. $d\sigma/dT$ tests the trace of $H_{\mu\nu}$, whereas $d\sigma_L/dT$, $d\sigma_T/dT$ and $d\sigma_I/dT$ are connected with other projections of $H_{\mu\nu}$.

The complete angular distribution can be written in the form

$$w(\theta,\chi) = 1 + \alpha(T)\cos^2\theta + \beta(T)\sin^2\theta\cos2\chi + \gamma(T)\sin2\theta\cos\chi \quad . \qquad (3.1.24)$$

Because of (3.1.23) $\beta(T)$ is related to $\alpha(T)$

$$\beta(T) = \frac{1}{4}\left[1-\alpha(T)\right] \quad . \qquad\qquad\qquad\qquad (3.1.25)$$

Of course, this would be an important test. $\alpha(T)$, $\beta(T)$ and $\gamma(T)$ are plotted as a function of T in Fig.3.6 /Koller, Sander, Walsh and Zerwas, 1980/. $\alpha(T)$ approaches 1 for T = 1, i.e. the angular distribution approaches the characteristic behaviour $1 + \cos^2\theta$ as for $q\bar{q}$ production. The deviations from the $1 + \cos^2\theta$ behaviour are most prominent for small T-values near the lower boundary T = 2/3. The cross section $d\sigma_L/dT$ etc. and similarly $\alpha(T)$ etc. depend on the quark-gluon coupling α_s. Therefore the measurement of these cross sections yields independent information on the value of α_s. The curves in Figs.3.4,5 are calculated for W = 30 GeV and Λ = 0.5 GeV with N_f = 5 (corresponds to α_s = 0.2).

The conspicuous and presumably the most easily measurable property of (3.1.24) is the deviation of the θ distribution from $(1 + \cos^2\theta)$ if T is decreased. Preliminary data of the JADE Collaboration have been reported by Elsen /1981/. For T < 0.95 (T < 0.90) he found $\alpha = 0.68 \pm 0.34$ ($\alpha = 0.50 \pm 0.47$). As we can see, these results are consistent with the curves in Fig.3.6.

44

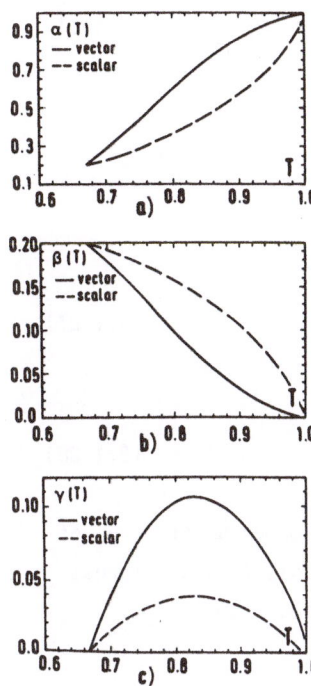

<u>Fig.3.6.</u> Plot of coefficients a) $\alpha(T)$, b) $\beta(T)$ and c) $\gamma(T)$ as a function of T for vector and scalar gluon theory in order α_s

More theoretical work about these polarization dependent cross sections is found in Pi, Jaffe and Low /1978/, Nandi and Wada /1980/, Bopp and Schiller /1980/, Johnson and Wu-ki Tung /1982/. If the interference with the Z pole is considered also additional terms appear in the angular distribution. The details can be found in Schierholz and Schiller /1979/, Laermann, Streng and Zerwas /1980/, Koller, Schiller and Wähner /1982/, Jeršak, Laermann and Zerwas /1982/.

3.1.4 $x_\perp$ Distributions

The thrust distribution is not the only single variable distribution considered. A similar distribution is the one using the transverse momentum $\sum_i |p_\perp^i|$ with respect to the thrust axis as remaining variable. Instead of $\sum_i |p_\perp^i|$ we take the scaled variable

$$x_\perp = \frac{\sum_i |p_\perp^i|}{W} \tag{3.1.26}$$

where i runs over the particles of the final state. This variable has the same advantage as thrust. Since it is linear in momentum it is least sensitive to infrared and clustering effects /Georgi and Machacek, 1977; Georgi and Sheiman, 1979/. For the $q\bar{q}g$ final state we have, if the thrust axis is along $\mathbf{p_1}$ (see Fig.3.7)

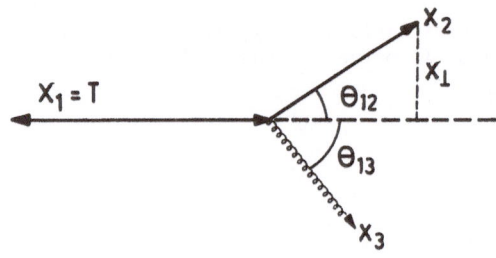

Fig.3.7. Definition of $x_\perp$ in $q\bar{q}g$ final state

$$x_\perp = x_2 \sin\theta_{12} = x_2 \sin\theta_{13} = \frac{2}{x_1} \left[(1-x_1)(1-x_2)(1-x_3) \right]^{1/2} \qquad (3.1.27)$$

since

$$\sin\theta_{ij} = \frac{2}{x_i x_j} \left[(1-x_1)(1-x_2)(1-x_3) \right]^{1/2} \qquad (3.1.28)$$

and θ_{ij} is the angle between the momenta $\mathbf{p}_i$ and $\mathbf{p}_j$ ($\mathbf{p}_1$ is the momentum of q, $\mathbf{p}_2$ of $\bar{q}$ and $\mathbf{p}_3$ of g) and similarly for the other two cases: thrust axis $\parallel \mathbf{p}_2$ and thrust axis $\parallel \mathbf{p}_3$. In (3.1.27) $x_1 = T$. The $x_\perp$ distribution is calculated from

$$\frac{1}{\sigma^{(2)}} \frac{d\sigma}{dx_\perp} = \frac{1}{\sigma^{(2)}} \int\limits_{T_{min}}^{T_{max}} dT \frac{d^2\sigma}{dT dx_\perp} \qquad (3.1.29)$$

where $d^2\sigma/dTdx_\perp$ must be calculated from (3.1.11) by transforming the variables x_1 and x_2 into T and $x_\perp$ in the appropriate kinematic regions I, II etc. in Fig.3.2. The limits of integration are $T_{max} = 1-x_\perp^2$ and T_{min}, the solution of

$$x_\perp^2 = \frac{4}{T_{min}^2} (1-T_{min})^2 (2T_{min}-1) \qquad (3.1.30)$$

where $2/3 \le T_{min} < T_{max} \le 1$.

Actually $x_\perp^2$ is nothing else than spherocity introduced in Sect.2.3. We have the relation

$$x_\perp^2 = \frac{\pi^2}{16} S' \quad . \qquad (3.1.31)$$

In general, the thrust and the spherocity axes need not be identical. However, they coincide for a three particle final state, both aligning with the momentum of the most energetic particle. The double differential cross section $d^2\sigma/dTdS'$ has been computed by De Rujula, Ellis, Floratos and Gaillard /1978/ which can easily be transformed into $d^2\sigma/dTdx_\perp^2$ with the result:

$$\frac{1}{\sigma^{(2)}} \frac{d^2\sigma}{dTdx_\perp^2} = \frac{\alpha_s}{2\pi} C_F \frac{T}{4(1-T)[1-x_\perp^2/(1-T)]^{1/2}} \left\{ 2\left[\frac{T^2+x_{2+}^2}{(1-T)(1-x_{2+})} + \frac{T^2+x_{2-}^2}{(1-T)(1-x_{2-})} \right] \right.$$

$$\left. + \frac{(2-T-x_{2+})^2+x_{2+}^2}{(T+x_{2+}-1)(1-x_{2+})} + \frac{(2-T-x_{2-})^2+x_{2-}^2}{(T+x_{2-}-1)(1-x_{2-})} \right\} \tag{3.1.32}$$

where

$$x_{2\pm} = 1 - \frac{1}{2} T \left(1 \pm \sqrt{1-x_\perp^2/(1-T)} \right) \quad . \tag{3.1.33}$$

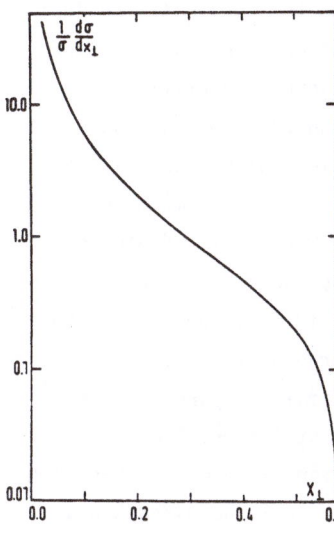

Fig.3.8. Distribution $(1/\sigma)d\sigma/dx_\perp$ as a function of $x_\perp$

$x_\perp$ varies in the interval $\theta \leq x_\perp \leq 1/\sqrt{3}$. The $x_\perp$ distribution is shown in Fig.3.8. Although $d\sigma/dx_\perp$ diverges as $x_\perp \to 0$, moments $\langle x_\perp^n \rangle$ of $x_\perp$ are finite. For example, for the moment of $x_\perp^2$ we obtain /De Rujula, Ellis, Floratos and Gaillard, 1978/

$$\langle x_\perp^2 \rangle = \frac{\alpha_s}{\pi} C_F \left(-\frac{229}{9} + 64 \ln\frac{3}{2} \right) \simeq 0.67 \frac{\alpha_s}{\pi} \quad . \tag{3.1.34}$$

So in lowest order QCD perturbation theory $\ln W^2 \langle x_\perp^n \rangle$ is independent of W^2 since

$$\alpha_s = \frac{12\pi}{33-2N_f} \ln\frac{W^2}{\Lambda^2}$$

in the same way as the moments of $(1-T)^n$ /Hoyer, Osland, Sander, Walsh and Zerwas, 1979/.

3.1.5 The Acollinearity Distributions

Another measurable quantity, which can be derived quite easily from the basic formulae in Sect.3.1.2, is the energy correlation as proposed by Basham, Brown, Ellis and Love /1978,1979/ (see also Fox and Wolfram /1980/). Experimentally two hadrons a and b are observed with energy fractions x_a and x_b, respectively, at an angle χ to each other. The energy correlation function is then

$$\frac{d\Sigma}{d\cos\chi} = \frac{1}{\sigma_{tot}} \sum_{a,b} \int dx_a dx_b \frac{x_a x_b}{4} \frac{d^3\sigma(e^+e^- \to abX)}{dx_a dx_b d\cos\chi} \quad . \tag{3.1.35}$$

Experimentally this is readily computed and has the advantage that no jet axis has to be determined as it was the case for $d\sigma/dT$ and $d\sigma/dx_\perp$. It is also easily calculable in perturbation theory, being infrared finite, at least for $\cos\chi \neq \pm 1$, in just the same way as $d\sigma/dT$ for $T \neq 1$ and $d\sigma/dx_\perp$ for $x_\perp \neq 0$. Moreover there was the hope that (3.1.35) would be less sensitive to fragmentation than other quantities. The reason for this was that in relating the correlation function (3.1.35) for hadrons to that for partons the energy weighted fragmentation function enters and this is normalized by energy conservation, since a parton must convert all its momentum into hadrons in the process of fragmentation (see for example Marquardt and Steiner /1980/). This gives a one-to-one relation between the energy correlation function for hadrons, i.e. (3.1.35), and that for partons, which is given by the same formula, except that a,b are partons now. But this argument is true only, if the transverse momentum fractions x_{T_i} in the fragmentation are very small compared to the energy fractions x_i. Apparently this is not the case, at least for PETRA energies, and strong dependence of the energy correlation on the fragmentation models has been established recently /Behrend et al., 1983/. But these topics will be discussed in detail later on. In this section we would like to present the energy correlations for the basic parton process of Fig.3.1b. For this purpose we take (3.1.11) and introduce instead of x_2 the variable $\zeta = (1/2)(1-\cos\theta_{12})$, where θ_{12} is the angle between quark and antiquark momentum. The relation is

$$x_2 = \frac{1-x_1}{1-x_1\zeta} \tag{3.1.36}$$

and the transformed cross section becomes

$$\frac{1}{\sigma^{(2)}} \frac{d^2\sigma}{d\cos\theta_{12} dx_1} = \frac{\alpha_s}{2\pi} C_F \frac{[(1-x_1)^2 + x_1^2(1-x_1\zeta)^2]}{2(1-\zeta)(1-x_1\zeta)^3} \quad . \tag{3.1.37}$$

In the same way we obtain for the cross section as a function of x_1 and $\zeta = (1/2)(1-\cos\theta_{13})$, where θ_{13} is the angle between quark and gluon momentum and

48

$$x_3 = \frac{1-x_1}{1-x_1\zeta} \tag{3.1.38}$$

the expression

$$\frac{1}{\sigma^{(2)}} \frac{d^2\sigma}{d\cos\theta_{13}dx_1} = \frac{\alpha_s}{2\pi} C_F \frac{\left[x_1^4\zeta^2(1-\zeta)^2+[1+2\zeta(1-\zeta)]x_1^2(1-\zeta x_1)^2+(1-\zeta x_1)^4\right]}{2\zeta(1-\zeta x_1)^3(1-x_1)} \quad . \tag{3.1.39}$$

The energy correlation function (3.1.35) for the correlations between quark, anti-quark and gluon follows from the integral

$$\frac{1}{\sigma^{(2)}} \frac{d\Sigma}{d\cos\chi} = \frac{1}{2\sigma^{(2)}} \int_0^1 dx_1 \left[x_1 x_2 \frac{d^2\sigma}{d\cos\theta_{12}dx_1} + 2x_1 x_3 \frac{d^2\sigma}{d\cos\theta_{13}dx_1}\right] \tag{3.1.40}$$

where $\cos\theta_{12}$ and $\cos\theta_{13}$ must be replaced by $\cos\chi = 1-2\zeta$ and x_2 and x_3 are substituted by (3.1.36) and (3.1.38)[1].

The result of the integration is

$$\frac{1}{\sigma^{(2)}} \frac{d\Sigma}{d\cos\chi} = \frac{\alpha_s}{8\pi} C_F \frac{3-2\zeta}{\zeta^5(1-\zeta)} [2(3-6\zeta+2\zeta^2)\ln(1-\zeta)+3\zeta(2-3\zeta)] \quad . \tag{3.1.41}$$

The energy correlations are singular for $\zeta \to 1$, which is $\chi \to \pi$ and $\zeta \to 0$, which is $\chi \to 0$. The leading singularity is for the opposite side correlations and has the form

$$\frac{1}{\sigma^{(2)}} \frac{d\Sigma}{d\cos\chi}_{\zeta\to 1} = \frac{\alpha_s}{4\pi} C_F \frac{1}{1-\zeta} \ln\frac{1}{1-\zeta} \tag{3.1.42}$$

/Brown and Ellis, 1981/.

The perturbative result (3.1.41) for the energy correlations is quite asymmetric about $\chi = \pi/2(\zeta = 1/2)$ as can be seen in Fig.3.9, where (3.1.41) has been plotted as a function of $\cos\chi$ for $\alpha_s = 0.20$. On the other hand the leading non-perturbative contributions caused by fragmentation in the process $e^+e^- \to q\bar{q}$ are expected to be even under the interchange $\chi \leftrightarrow \pi-\chi$ ($\zeta \leftrightarrow 1-\zeta$). Thus, as the energy increases, the perturbative QCD contribution (3.1.41) is expected to dominate in

$$AS(\cos\chi) = \frac{1}{\sigma^{(2)}} \left[\frac{d\Sigma(-\cos\chi)}{d\cos\chi} - \frac{d\Sigma(\cos\chi)}{d\cos\chi}\right] \tag{3.1.43}$$

with the non-perturbative corrections being of order $1/W$. $AS(\cos\chi)$ is plotted in Fig.3.10. It is large for $\chi \leq 50°$.

---

[1] The angle χ defined here should not be confused with the azimuthal angle in Sect. 3.1.2.

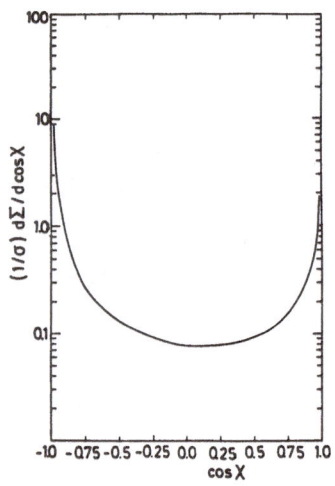
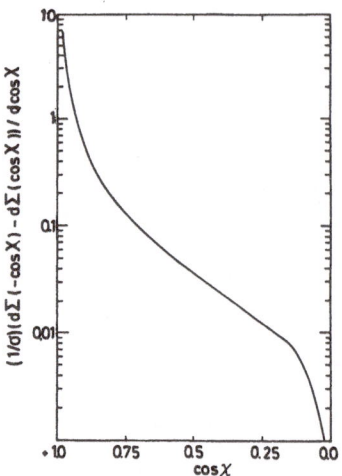

Fig.3.9. The $O(\alpha_s)$ cross section $(1/\sigma)d\Sigma/d\cos\chi$ for $e^+e^- \to q\bar{q}g$ as a function of $\cos\chi$ for $\alpha_s = 0.2$

Fig.3.10. The $O(\alpha_s)$ asymmetry cross section for $e^+e^- \to q\bar{q}g$ as a function of $\cos\chi$ for $\alpha_s = 0.2$

Concerning the non-perturbative contribution one must keep in mind that (3.1.41) includes contributions where the parton, which is not detected, becomes collinear with one of the detected partons. This is actually part of the 2-jet region and should be excluded from the integrations in (3.1.40) (near the upper integration point $x_1 = 1$). This would make the result cut-off dependent and much more compli- cated. When comparing to experimental data, this must be kept in mind (see Ali and Barreiro /1982/).

The expression (3.1.41) represents only the contribution to the energy correla- tions which results after integration over the angles θ and φ of the hadron plane with respect to the beam axis. The full cross section has the form (3.1.9). The other coefficients in (3.1.9) can be found in the work of Basham, Brown, Ellis and Love /1978,1979/. As a partial result we quote the distribution obtained after integra- tion over φ [$\varphi \equiv \chi$ in (3.1.9)]:

$$\frac{1}{\sigma^{(2)}} \frac{d^2\Sigma}{d\cos\chi d\cos\theta} = \frac{\alpha_s}{8\pi} C_F \frac{3}{2} [A(\zeta) + \cos^2\theta \cdot B(\zeta)] \tag{3.1.44}$$

where

$$A(\zeta) = \frac{1}{4(1-\zeta)} \left[\left(\frac{18}{\zeta^5} - \frac{42}{\zeta^4} + \frac{22}{\zeta^3}\right)\ln(1-\zeta) + \frac{18}{\zeta^4} - \frac{33}{\zeta^3} + \frac{7}{\zeta^2} + \frac{3}{\zeta} + 2 \right]$$

$$\tag{3.1.45}$$

$$B(\zeta) = \frac{1}{4(1-\zeta)} \left[\left(\frac{18}{\zeta^5} - \frac{66}{\zeta^4} + \frac{78}{\zeta^3} - \frac{32}{\zeta^2}\right)\ln(1-\zeta) + \frac{18}{\zeta^4} - \frac{57}{\zeta^3} + \frac{51}{\zeta^2} - \frac{9}{\zeta} - 6 \right] \quad .$$

Further work appeared in Soper /1983/, Brown and Li /1982/.

50

3.1.6 Influence of Beam Polarization

Up to now we considered cross sections for $e^+e^- \to q\bar{q}g$ only with unpolarized e^+ and e^- beams. One may ask whether polarized e^+ and e^- beams yield additional information.

To answer this question we express the differential cross section for $e^+e^- \to q\bar{q}g$ as a product of the lepton tensor $L_{\mu\nu}$ and the hadron tensor $H_{\mu\nu}$ introduced in Sect. 3.1.2:

$$d\sigma = L^{\mu\nu}H_{\mu\nu}dP \qquad (3.1.46)$$

where dP denotes the phase space element. We split (3.1.46) into contributions from the symmetric and antisymmetric lepton and hadron tensor combinations

$$d\sigma = \left(L^{\{\mu\nu\}}H_{\{\mu\nu\}} + L^{[\mu\nu]}H_{[\mu\nu]}\right)dP \qquad . \qquad (3.1.47)$$

In the one-photon approximation the antisymmetric lepton tensor can be seen to receive contributions only from longitudinally polarized beams. One has

$$L_{[\mu\nu]} = \frac{2im}{q^2}\,\varepsilon_{\mu\nu\rho\sigma}[S^{(+)}+S^{(-)}]^\rho q^\sigma \qquad (3.1.48)$$

where $q = p_+ + p_-$ and $S^{(+)}$ and $S^{(-)}$ are covariant polarization vectors of the positron and electron beams, respectively. In the center-of-mass system it has the simple form

$$L_{[ij]} = i\varepsilon_{ijk}v^k\left[\xi_z^{(+)}+\xi_z^{(-)}\right] \qquad (3.1.49)$$

with $\xi_z^{(\pm)}$ denoting the degree of longitudinal polarization of the electron and positron beams and $\mathbf{v}$ is a unit vector in the electron beam direction. Thus to detect the antisymmetric part $H_{[\mu\nu]}$ one needs longitudinal polarized $e^\pm$ beams with $\xi_z^{(+)} + \xi_z^{(-)} \neq 0$.

Quite generally, the hadron tensor depends on 5 structure function H_i (i = 1 ... 5); 4 are proportional to a symmetric and 1 to an antisymmetric tensor:

$$H_{\mu\nu} = (q^2 g_{\mu\nu}-q_\mu q_\nu)H_1 + \hat{p}_{1\mu}\hat{p}_{1\nu}H_2 + \hat{p}_{2\mu}\hat{p}_{2\nu}H_3$$

$$+ (\hat{p}_{1\mu}\hat{p}_{2\nu}+\hat{p}_{2\mu}\hat{p}_{1\nu})H_4 + (\hat{p}_{1\mu}\hat{p}_{2\nu}-\hat{p}_{2\mu}\hat{p}_{1\nu})H_5 \qquad (3.1.50)$$

where $\hat{p}_\mu = q^2 p_\mu - p \cdot q q_\mu$. This decomposition follows readily from gauge invariance $q^\mu H_{\mu\nu} = 0$ and the fact that only three independent four vectors are at our disposal (we have chosen q, p_1 and p_2). Since we have $H_{\mu\nu} = H^*_{\nu\mu}$ from the hermiticity of the electromagnetic current, we conclude that $H_1 \ldots H_4$ are real and H_5 is imaginary. Thus, H_5 is the only hadronic structure function that contributes to $L_{[\mu\nu]}H^{[\mu\nu]}$.

In order α_s the hadron tensor $H_{\mu\nu}$ is purely real. Therefore in this approximation $H_5 = 0$ (as can be seen also by inspecting (3.1.5)) and there is no effect from $L_{[\mu\nu]} \neq 0$. H_5 appears only in the approximation $O(\alpha_s^2)$ and higher. One needs final state interactions between q, $\bar{q}$ and g in order to obtain a non-real contribution to the q$\bar{q}$g production amplitude. The remaining four amplitudes in the symmetric part of $H_{\mu\nu}$ correspond to the four cross sections $d\sigma_U$, $d\sigma_L$, $d\sigma_T$ and $d\sigma_I$ introduced in Sect.3.1.2. They all can be determined from measurements of the angular distribution (3.1.8) with respect to the beam axis. Therefore, transversal polarization of the incoming beams does not give any information in addition to that which can be obtained from the angular distribution (3.1.9) already. This fact is well known and is true for all final states /Avram and Schiller, 1974/. Nevertheless, with transversal polarized beams it is much easier to measure the coefficients $\alpha(T)$, $\beta(T)$ and $\gamma(T)$ in the angular distribution (3.1.24) than without polarization. This is analogous to the measurement of the jet axis angular distribution at low energies with the SPEAR ring considered in Sect.2.2.

It is well known that electrons and positrons tend to polarize themselves in the direction of the magnetic fields. Recent experiments at PETRA have shown that it is possible to circumvent depolarizing resonances and reasonable transversal polarization can be achieved. With transverse polarization at our disposal it should be possible to measure all sorts of angular correlation coefficients. Having $e^{\pm}$ beams with general polarization, transverse or/and longitudinal the differential cross section for $e^+e^- \to q\bar{q}g$ has the following from (Hirshfeld, Kramer and Schiller /1974/, in this reference the formulae were derived for a general 3-particle final state, see also Avram and Schiller /1974/):

$$
\frac{(2\pi)^2 d^5\sigma}{d\cos\theta d\chi d\phi dx_1 dx_2} = \frac{3}{8}\left[(1+Z)(1+\cos^2\theta)+X(\phi)\sin^2\theta\right]\frac{d^2\sigma_U}{dx_1 dx_2} + \frac{3}{4}\left[1+Z-X(\phi)\right]\sin^2\theta\frac{d^2\sigma_L}{dx_1 dx_2}
$$

$$
+ \frac{3}{4}\left\{\left[(1+Z)\sin^2\theta+X(\phi)(1+\cos^2\theta)\right]\cos2\chi+2Y(\phi)\cos\theta\sin2\chi\right\}\frac{d^2\sigma_T}{dx_1 dx_2}
$$

$$
- \frac{3}{\sqrt{2}}\left\{\left[1+Z-X(\phi)\right]\cos\theta\cos\chi-Y(\phi)\sin\chi\right\}\sin\theta\frac{d^2\sigma_I}{dx_1 dx_2} + \frac{3}{\sqrt{2}}L\sin\theta\sin\chi\frac{d^2\sigma_H}{dx_1 dx_2} \quad . \quad (3.1.51)
$$

The quantities X, Y, Z and L depend on the polarization parameters $\xi_{x,y,z}^{(\pm)}$

$$X(\phi) = \left[\xi_x^{(+)}\xi_x^{(-)} - \xi_y^{(+)}\xi_y^{(-)}\right]\cos2\phi + \left[\xi_x^{(+)}\xi_y^{(-)} + \xi_y^{(+)}\xi_x^{(-)}\right]\sin2\phi$$

$$Y(\phi) = \left[\xi_x^{(+)}\xi_y^{(-)} + \xi_y^{(+)}\xi_x^{(-)}\right]\cos2\phi - \left[\xi_x^{(+)}\xi_x^{(-)} - \xi_y^{(+)}\xi_y^{(-)}\right]\sin2\phi$$

$$Z \quad = \xi_z^{(+)} \cdot \xi_z^{(-)}$$

$$L \quad = \xi_z^{(+)} + \xi_z^{(-)} \tag{3.1.52}$$

$\xi^{(\pm)}$ are the $e^\pm$ polarization vectors in the corresponding rest frames, i.e. $\xi_x^{(\pm)}$ and $\xi_y^{(\pm)}$ mean transverse and $\xi_z^{(\pm)}$ longitudinal beam polarization. The labeling of the cross sections $d\sigma_k$ (k = U, L, T, I) has been explained already. $d\sigma_H$ is the imaginary part of the interference between transversely and longitudinally polarized virtual photons:

$$\frac{d^2\sigma_H}{dx_1dx_2} \sim \text{Im } \varepsilon_\alpha(+1) H_{\alpha\beta} \varepsilon_\beta(0)^*, \quad \alpha,\beta = 1,2,3 \quad . \tag{3.1.53}$$

This cross section is proportional to H_5 in (3.1.50). As we mentioned already this can appear only in higher order perturbative contributions. In lowest non-trivial order $O(\alpha_s)$ the cross section has the form (3.1.51) with $d^2\sigma_H/dx_1dx_2 = 0$. The cross section $d^2\sigma_k/dx_1dx_2$ (k = U, L, T, I) in $O(\alpha_s)$ can be found in (3.1.11-15) for three cases concerning the definition of coordinates axes in terms of final state momenta.

Actually α_H is particularly interesting for testing QCD since it contributes to order α_s^2 and higher only, being proportional to the imaginary part of the hadronic tensor, it must involve at least one loop. Unfortunately σ_H is zero for massless quarks /Körner, Kramer, Schierholz, Fabricius and Schmidt, 1980/. But a sizable cross section has been predicted for heavy quarks by Fabricius, Schmitt, Kramer and Schierholz /1980/. Their result was expressed in a special coordinate system which uses the normal to the event plane as the z direction for the hadron system with the special requirement that the z direction points into the direction $\mathbf{p}_1 \times \mathbf{p}_2$ if $x_1 > x_2$ and $\mathbf{p}_2 \times \mathbf{p}_1$ if $x_2 > x_1$, respectively. This convention has been chosen since it should be relatively easy to locate the heavy quark (antiquark) jets. Since $\sin\theta\sin\chi = \cos\eta$, η being the polar angle of the z direction and the electron beam direction (Fig.3.11) (3.1.51) reduces for purely longitudinally polarized beams to:

$$\frac{d^3\sigma}{d\cos\eta dx_1 dx_2} = \frac{3}{8}(1+Z)(1+\frac{1}{2}\sin^2\eta)\frac{d^2\sigma_U}{dx_1dx_2} + \frac{3}{4}(1+Z)(1-\frac{1}{2}\sin^2\eta)\frac{d^2\sigma_L}{dx_1dx_2}$$

$$+ \frac{3}{4}(1+Z)(-1+\frac{3}{2}\sin^2\eta)\frac{d^2\sigma_T}{dx_1dx_2} + \frac{3}{\sqrt{2}}L\cos\eta\frac{d^2\sigma_H}{dx_1dx_2} \tag{3.1.54}$$

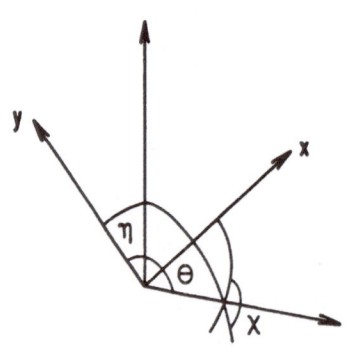

e⁻ beam direction

Fig.3.11. Three-jet kinematics. The event lies in the $\overline{(x,z)}$ plane. The convention for the angles θ, χ and η is as follows: $\overrightarrow{Oz}$ is along the direction of the fastest of either quark or antiquark jet, and the second most energetic quark/antiquark jet lies in the half-plane $x > 0$. The normal $\overrightarrow{Oy}$ to the event plane is in the direction sign $(x_1-x_2)(p_1 \times p_2)$. $0 \leq \theta \leq \pi$, $0 \leq \chi \leq 2\pi$ and $0 \leq \eta \leq \pi$.

and σ_I drops out. The asymmetry in $\cos\eta$ which is proportional to σ_H has been calculated in the form of a quantity R defined by

$$R = \frac{8}{\sqrt{2}} \frac{d^2\sigma_H}{dx_1 dx_2} \left/ \frac{d^2\sigma}{dx_1 dx_2} \right. \tag{3.1.55}$$

with the denominator approximated by the dominant lowest order cross section. The result for $m/\sqrt{q^2} = 0.25$, where m is the quark mass, is shown in Fig.3.12. The coupling constant is $\alpha_s = 0.2$. Instead of the quark energies x_1 and x_2 the variables are thrust T

$$T = \max\left(\sqrt{x_1^2 - 4m^2/q^2}, \sqrt{x_2^2 - 4m^2/q^2}, x_3\right) \frac{2}{\sqrt{x_1^2 - 4m^2/q^2} + \sqrt{x_2^2 - 4m^2/q^2}} \tag{3.1.56}$$

$x_3 = 2E_3/\sqrt{q^2}$, $x_1 + x_2 + x_3 = 2$ and

$$\cos\theta_{12} = \left(x_1 x_2 - 2x_1 - 2x_2 + \frac{4m^2}{q^2}\right) \left/ \left[\left(x_1^2 - \frac{4m^2}{q^2}\right)\left(x_2^2 - \frac{4m^2}{q^2}\right)\right]^{1/2}\right. \quad .$$

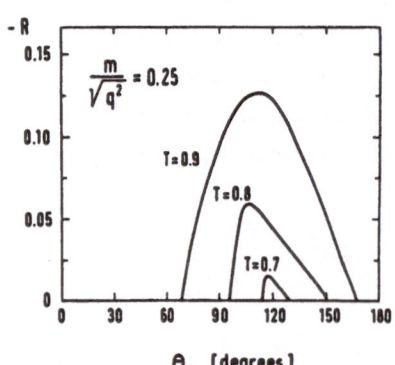

θ_{12} [degrees]

Fig.3.12. The asymmetry parameter R for $m/\sqrt{q^2} = 0.25$ and various thrust values T as a function of θ_{12}

θ_{12} being the angle between the quark and antiquark momentum to fix the 3-jet kinematics which are easier to determine experimentally. The predicted asymmetry depends quite strongly on the value of $m/\sqrt{q^2}$ as is to be expected. For the yet to be discovered top quark the effect should be measurable. For the bottom quark ($m \simeq 5$ GeV) and 40 GeV center-of-mass energy the effect will still be on the percent level.

According to the two classes of diagrams contributing to σ_H, those involving the triple-gluon coupling and the QED type diagrams (see Sect.3.2) R can be written

$$R = \frac{1}{N_c} \text{Tr} \left(i f_{\alpha\beta\gamma} \frac{\lambda_\beta}{2} \frac{\lambda_\gamma}{2} \frac{\lambda_\alpha}{2} \right) r_C + \frac{1}{N_c} \text{Tr} \left(\frac{\lambda_\beta}{2} \frac{\lambda_\alpha}{2} \frac{\lambda_\beta}{2} \frac{\lambda_\alpha}{2} \right) r_E \quad . \tag{3.1.57}$$

It was found that, apart from the edges of phase space, where R becomes very small, $r_C \simeq r_E$. Noticing that the first trace in (3.1.57) is equal to (-2) and the second equal to (-2/9) one obtains with the coupling constant relation which leads to equal $q\bar{q}g$ integrated cross sections $\alpha_A = (4/3) \alpha_s$ a relation between the R in an abelian vector gluon theory (R_{QAD}) and the R for QCD:

$$R_{QAD} \simeq - \frac{4}{5} R_{QCD} \quad . \tag{3.1.58}$$

Thus R is a measure of the gluonic self-interactions. In order to prove QCD it would be sufficient to establish the sign of R.

3.1.7 Jet Multiplicities and the Total Cross Section

We learned in the previous sections that the 3-jet cross section is well defined outside the singular region $x_1, x_2 = 1$. The same is true for the other cross sections σ_U, σ_L, σ_T, σ_I and σ_H. One may wonder how $d^2\sigma/dx_1 dx_2$ should be interpreted in such kinematical regions which include the singular regions $x_1 = 1$ and/or $x_2 = 1$.

In the direct vicinity of $x_1, x_2 = 1$ the $q\bar{q}g$ final state is essentially a 2-jet configuration, for which $x_1 = x_2 = 1$. For $x_1 = 1$ or $x_2 = 1$ the quark or antiquark momenta are collinear with the gluon momenta, so being equivalent to one parton. For $x_1 = x_2 = 1$ we have the emission of a zero energy gluon, so again a 2-jet configuration evolves. The situation is familiar from QED. Take, for example, the reaction $e^+e^- \rightarrow \mu^+\mu^-\gamma$ with a very soft photon in the final state. Here also, this final state is indistinguishable from a pure $\mu^+\mu^-$ state. In the same way we must consider the $q\bar{q}g$ events with $x_1, x_2 \simeq 1$ as 2-jet events and must consider them together with the pure $q\bar{q}$ events. For this we need a criterion which defines the 2-jet region in the $q\bar{q}g$ phase space. Such a criterion is more or less arbitrary. It must be a region of the $q\bar{q}g$ phase space (see Fig.3.2) which includes fully the lines $x_1, x_2 = 1$. The most simple boundary is

$$1-y \leq x_1 \leq 1 \quad , \qquad 0 \leq x_2 \leq 1$$

$$1-y \leq x_2 \leq 1 \quad , \qquad 0 \leq x_1 \leq 1-y \quad .$$

(3.1.59)

Instead of the variables x_1, x_2 it is more convenient to use squared invariant masses of two parton pairs, i.e.

$$y_{ij} = (p_i + p_j)^2 / q^2 = 2 p_i p_j / q^2 \quad .$$

(3.1.60)

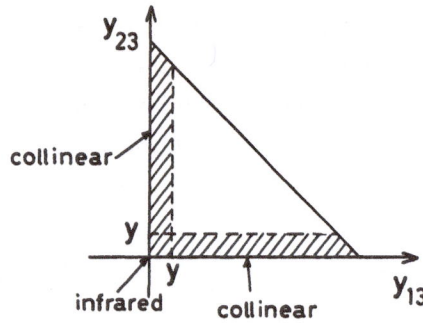

Fig.3.13. $q\bar{q}g$ phase space expressed in terms of y_{13} and y_{23} with boundaries for 2-jet regions

Expressed in $y_{13} = 1-x_2, y_{23} = 1-x_1$ ($y_{13}+y_{23}+y_{12} = 1$) the 2-jet phase space (3.1.59) is shown in Fig.3.13. Of course $y < 1/2$ and we choose $y \ll 1$. This yQ^2 is the upper limit of the invariant masses squared $(p_1+p_3)^2$ and $(p_2+p_3)^2$ formed with the quark and gluon and antiquark and gluon momenta, respectively. Here the boundary value y serves the same purpose as the familiar ΔE in QED for defining the infrared photon region. The area near the origin in the y_{13}, y_{23} plane ($y_{13}, y_{23} \geq 0$, $1-y_{13}-y_{23} \geq 0$) is the infrared region, whereas the strips in the vicinity of y_{13} or $y_{23} = 0$ are the regions of collinear singularities.

Of course, the integral of $d^2\sigma/dx_1 dx_2$ over the region (3.1.59), i.e. over the 2-jet region of the $q\bar{q}g$ phase space, diverges because of the infrared and collinear singularities. However, this part of integrated cross section alone has no physical meaning. It must be taken together with all other 2-jet contributions. Only the total 2-jet cross section is a measurable quantity. Other 2-jet contributions in $O(\alpha_s)$ are the virtual one-loop corrections to $e^+e^- \to q\bar{q}$ shown in Fig.3.1a. Only the sum of these two contributions, the virtual corrections and the integral of $q\bar{q}g$ over the strips in Fig.3.13 together gives a finite 2-jet cross section.

To calculate this sum we proceed as follows. As mentioned already in Sect.3.1.1 we introduce the arbitrary dimension $n = 4-2\varepsilon$ in the integral to control the divergences in the integral over the $q\bar{q}g$ differential cross section. This differential cross section is in n dimensions

$$\frac{d^2\sigma}{dy_{13}dy_{23}} = \sigma^{(2)} \left(\frac{4\pi\mu^2}{q^2}\right)^\epsilon \frac{1}{\Gamma(1-\epsilon)} \frac{\alpha_s(\mu^2)}{2\pi} C_F (y_{13}y_{23}y_{12})^{-\epsilon} B(y_{13},y_{23}) \tag{3.1.61}$$

where

$$B(y_{23},y_{23}) = B^V(y_{13},y_{23}) + \epsilon B^S(y_{13},y_{23}) \tag{3.1.62}$$

with

$$B^V(y_{13},y_{23}) = \frac{y_{13}}{y_{23}} + \frac{y_{23}}{y_{13}} + \frac{2y_{12}}{y_{13}y_{23}} \tag{3.1.63}$$

$$B^S(y_{13},y_{23}) = \frac{y_{13}}{y_{23}} + \frac{y_{23}}{y_{13}} + 2 \tag{3.1.64}$$

and

$$\sigma^{(2)} = \sigma^{(2)}\Big|_{\epsilon=0} \left(\frac{4\pi\mu^2}{q^2}\right)^\epsilon \frac{\Gamma(2-\epsilon)}{\Gamma(2-2\epsilon)} \quad . \tag{3.1.65}$$

$\sigma^{(2)}\Big|_{\epsilon=0} = 0$ is given in (3.1.12). μ is an arbitrary mass, introduced to make the coupling $g = (4\pi\alpha_s)^{1/2}$ dimensionless in n dimensions. Of course, (3.1.61) goes over into (3.1.11) in the limit $n \to 4$ ($\epsilon \to 0$).

The integration of (3.1.61) over the 2-jet region (3.1.59) yields the contribution of the bremsstrahlungsgraphs to the 2-jet cross section

$$\sigma^{2\text{-jet}}(q\bar{q}g) = \sigma^{(2)} \left(\frac{4\pi\mu^2}{q^2}\right)^\epsilon \frac{\alpha_s(\mu^2)}{2\pi} C_F \frac{\Gamma(1-\epsilon)}{\Gamma(1-2\epsilon)}$$

$$\cdot \left(\frac{2}{\epsilon^2} + \frac{3}{\epsilon} - 2\ln^2 y - 3\ln y + 4y\ln y - \frac{\pi^2}{3} + 7\right) \quad . \tag{3.1.66}$$

The interference of the virtual diagrams in Fig.3.1a with the zeroth order diagrams in Fig.2.1 gives us the $O(\alpha_s)$ contribution of the virtual diagrams to $\sigma^{2\text{-jet}}$

$$\sigma^{2\text{-jet}}(q\bar{q}) = \sigma^{(2)} \left(\frac{4\pi\mu^2}{q^2}\right)^\epsilon \frac{\alpha_s(\mu^2)}{2\pi} C_F \frac{\Gamma(1-\epsilon)}{\Gamma(1-2\epsilon)} \left(-\frac{2}{\epsilon^2} - \frac{3}{\epsilon} + \frac{2\pi^2}{3} - 8\right) \quad . \tag{3.1.67}$$

In the dimensional regularization method of t'Hooft and Veltman /1972/ the infrared and collinear divergences appear as poles in $\epsilon = (4-n)/2(\epsilon^{-2}$ is the infrared divergence, ϵ^{-1} are infrared and collinear divergences) which cancel in the sum of (3.1.66) and (3.1.67) so that the limit $\varepsilon \to 0$ can be taken. This sum (for $\varepsilon \to 0$) is the $O(\alpha_s)$ contribution to the 2-jet cross section /Kramer, 1982/:

$$\sigma^{2\text{-jet}}(y) = \sigma^{(2)} \left[1 + \frac{\alpha_s(\mu^2)}{2\pi} C_F \left(-2\ln^2 y - 3\ln y + 4y\ln y - 1 + \frac{\pi^2}{3} \right) \right] \quad . \tag{3.1.68}$$

In (3.1.68) we included the zeroth order contribution $\sigma^{(2)}$. In $O(\alpha_s)$ $\sigma^{2\text{-jet}}$ depends now on y. It is important to notice that in higher order QCD the 2-jet cross section can be defined only on the basis of a parameter, in this case y, which defines the separation of 2- and 3-jet events. This parameter, which has the function of the ΔE for photons in QED must be chosen in accordance with the parameter used experimentally to separate 2- and 3-jet events. The necessity to fix such a parameter for the computation of $\sigma^{2\text{-jet}}$ in QCD is closely connected with the infrared singular behaviour due to massless gluons and quarks.

We remark that in the formula (3.1.68) all terms $O(y)$ except such $\sim y\ln y$ were neglected, so that (3.1.68) can be applied only for $y \ll 1$.

Our result (3.1.68) is in agreement with the so-called Kinoshita-Lee-Nauenberg theorem /Kinoshita, 1960; Lee and Nauenberg, 1966/. This theorem which was originally stated for QED processes says that cross sections for $e^+e^- \to X$ defined in such a way that the final state X is a sum over all indistinguishable configurations are finite. In our case the indistinguishable configurations are the states $q\bar{q}$ and $q\bar{q}g$ inside the strips of Fig.3.13.

The integration of $d^2\sigma/dy_{13}dy_{23}$ over $q\bar{q}g$ phase space complementary to (3.1.59), i.e. the triangle without the strips in Fig.3.13, yields the integrated 3-jet cross section $\sigma^{3\text{-jet}}(y)$

$$\sigma^{3\text{-jet}}(y) = \sigma^{(2)} \frac{\alpha_s}{2\pi} C_F \left(2\ln^2 y + 3\ln y - 4y\ln y + \frac{5}{2} - \frac{\pi^2}{3} \right)^2 \quad . \tag{3.1.69}$$

This result is obtained by integrating (3.1.60) with $\varepsilon = 0$, the infrared regularization is not needed. But it is important to notice that $\sigma^{3\text{-jet}}$ will always depend on parameters necessary to separate the 2- and 3-jet region, i.e. y in our case.

The sum of $\sigma^{2\text{-jet}}(y)$ and $\sigma^{3\text{-jet}}(y)$ is the total cross section

$$\sigma_{tot} = \sigma^{(2)} \left(1 + \frac{\alpha_s}{2\pi} \frac{3C_F}{2} \right) \quad . \tag{3.1.70}$$

σ_{tot} must be independent of y. (3.1.70) is in accordance with the other part of the Kinoshita-Lee-Nauenberg theorem stating that the fully inclusive e^+e^- cross section is finite in the limit of vanishing quark masses (free of mass singularities). (3.1.70) gives us the total e^+e^- annihilation cross section into hadrons in the order α_s, equivalent with the well-known formula for R introduced in Sect.2.1

$$R = \left(3 \sum_f Q_f^2\right)\left(1 + \frac{\alpha_s}{\pi}\right) \quad . \tag{3.1.71}$$

$\sigma^{2\text{-jet}}(y)/\sigma_{tot}$ and $\sigma^{3\text{-jet}}(y)/\sigma_{tot}$ are the 2-jet and 3-jet multiplicities up to $O(\alpha_s)$ in perturbation theory. They also depend on y which must be chosen in agreement with the parameter chosen to divide 2- and 3-jet events in the experimental data.

It is clear that our choice of the squared invariant jet mass for distinguishing 2 and 3 jets is not the only one possible. Another choice is the parametrization introduced by Sterman and Weinberg /1977/. They define a 2-jet event in the $q\bar{q}g$ phase space if the following conditions are satisfied

(i) $0 \le x_3 \le \varepsilon$

or $\tag{3.1.72}$

(ii) $\varepsilon \le x_3 \le 1$ and $\theta_{13} \le \delta$ or $\theta_{23} \le \delta$.

$\theta_{13}(\theta_{23})$ are the angles between quark (antiquark) and gluon momentum. Thus we have a contribution to $\sigma^{2\text{-jet}}(\varepsilon,\delta)$ if the gluon is emitted in a cone of angle δ with respect to the quark or antiquark momentum or with an energy $p_3 \le \varepsilon W/2$ outside the cones. The 2-jet cross section due to this definition was first calculated by Sterman and Weinberg /1977/ and is

$$\sigma^{2\text{-jet}}(\varepsilon,\delta) = \sigma^{(2)}\left[1 + \frac{\alpha_s(\mu^2)}{2\pi} C_F\left(-4\ln\varepsilon\cdot\ln\frac{\delta^2}{4} - 3\ln\frac{\delta^2}{4} - \frac{2\pi^2}{3} + 5\right)\right] \quad . \tag{3.1.73}$$

This formula is correct only for small ε and δ. Terms proportional to ε and δ were neglected. The complete formula has been computed by Binetruy and Girardi /1979/, by Stevenson /1978/ and by Weeks /1979/. From the difference $\sigma_{tot} - \sigma^{2\text{-jet}}(\varepsilon,\delta)$ one obtains $\sigma^{3\text{-jet}}(\varepsilon,\delta)$. We notice that (3.1.68) and (3.1.73) agree in the leading terms if we identify $\varepsilon^2 = \delta^2/4 = y$.

Which values should be chosen for y or ε,δ to have a reasonable perturbation theory? From (3.1.68) we see that $\sigma^{2\text{-jet}}(y)$ decreases with decreasing y and may become negative for small enough y. But $\sigma^{2\text{-jet}}(y)$ being a physical cross section must be positive. This shows that y must be reasonably large; how large, depends on the value of α_s. Similarly $\sigma^{3\text{-jet}}(y)$ increases like $\ln^2 y$ with decreasing y and may become larger than $\sigma^{(2)}(1 + \alpha_s/\pi)$ for small enough y. This is also unphysical. So let us define

$$\sigma_{red}^{2\text{-jet}}(y) = \sigma^{2\text{-jet}}(y)/\sigma^{(2)} \tag{3.1.74}$$

and similarly $\sigma_{red}^{3\text{-jet}}(y)$. Then

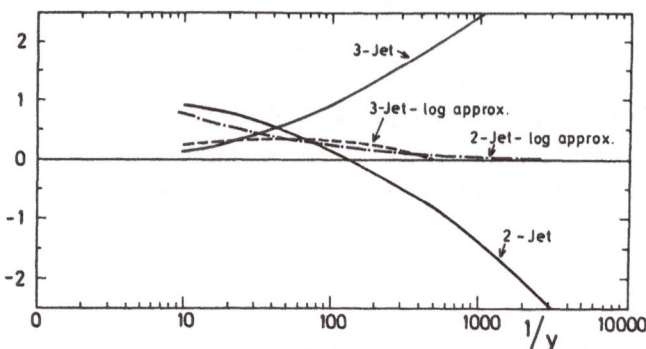

Fig.3.14. 2-jet and 3-jet cross section as a function of y in units of $\sigma^{(2)}$ for $\alpha_s/\pi = 0.05$. Dividing by $(1 + \alpha_s/\pi)$ yields jet multiplicities. Dashed (dashed-dotted) curve gives 3-jet (2-jet) cross section divided by $\sigma^{(2)}$ in leading-log approximation summed to all orders

$$\sigma_{red}^{2\text{-jet}}(y) + \sigma_{red}^{3\text{-jet}}(y) = 1 + \frac{\alpha_s}{\pi} \tag{3.1.75}$$

so that, up to a factor $(1 + \alpha_s/\pi)^{-1}$, $\sigma_{red}^{2\text{-jet}}(y)$ and $\sigma_{red}^{3\text{-jet}}(y)$ are equal to the 2- and 3-jet multiplicities. These quantities are plotted in Fig.3.14 as a function of y^{-1} for $\alpha_s/\pi = 0.05$. This is a reasonable value for α_s in agreement with experimental determinations which will be discussed in the next section. Then it follows from the graphs in Fig.3.14 that $\sigma_{red}^{3\text{-jet}}(y) < 1$ if $y > 0.01$. Of course, to have a reasonable perturbation theory, we require $\sigma_{red}^{3\text{-jet}}(y) < 0.4$. For this we need $y > 0.03$. On the other hand y should not be too large since terms proportional to y and $y^2 \ln^2 y$ etc. were neglected in (3.1.68) and $\sigma^{3\text{-jet}}(y)$ should not be too small if one wants to test the details of 3-jet events. Therefore a reasonable range is $0.03 \leq y \leq 0.05$. For y = 0.05 we have $\sigma_{red}^{3\text{-jet}} = 0.29$, so that $\sigma^{3\text{-jet}}/\sigma_{tot} = 0.28$ and $\sigma^{2\text{-jet}}/\sigma_{tot} = 0.72$ which is a quite reasonable choice for y. This choice corresponds roughly to $\epsilon = 0.2$, $\delta^2/4 = 0.1$ in the Sterman-Weinberg definition and yields

$$\sigma^{3\text{-jet}}/\sigma_{tot} = 0.35 \quad \text{and} \quad \sigma^{2\text{-jet}}/\sigma_{tot} = 0.65 \quad .$$

$T_0 = 1-y$ with y = 0.05 is also the most reasonable boundary for single variable thrust distributions. This means that the integration of the thrust distribution (3.1.18) in the internal $2/3 \leq T \leq T_0$ yields, up to terms of order $y = 1-T_0$, the same integrated 3-jet cross section as (3.1.69).

$T_0 = 0.95$ was also used in the phenomenological models of Hoyer et al. /1979/ and Ali et al. /1980/ for separating 2 and 3 jets in the $q\bar{q}g$ intermediate state before fragmentation.

From this viewpoint the mean value of (1-T) defined in (3.1.20) or of $x_\perp^2$ in (3.1.34) should also be defined with a cut-off. This means <1-T>, for example, must be defined by

$$<1-T>_{T_0} = \frac{1}{\sigma_{tot}} \int\limits_{2/3}^{T_0} dT \frac{d\sigma}{dT} (1-T) \quad .$$

(3.1.76)

This amounts to $<1-T>_{T_0} = 0.59(\alpha_s/\pi)(1+\alpha_s/\pi)^{-1}$ for $T_0 = 0.95$ instead of $1.05(\alpha_s/\pi)(1+\alpha_s/\pi)^{-1}$ for $T = 1$ [see (3.1.21)]. This way we have an appreciable reduction of the perturbative contribution to <1-T>. Therefore (3.1.20) was the limit of (3.1.75) for $T_0 \to 1$, which, however, does not give a reasonable estimate of the perturbative <1-T> [compared to (3.1.20) we used σ_{tot} instead of $\sigma^{(2)}$ for normalization in (3.1.76)]. Of course, the remaining contribution of <1-T> is non-perturbative and can be calculated only with the help of phenomenological fragmentation models. This replaces <1-T> for 2 jets which is zero in perturbation theory. For a somewhat different point of view, see Ch. Berger et al. /1982/. The formulae (3.1.68) and (3.1.69) which we obtained for $\sigma^{2-jet}(y)$ and $\sigma^{3-jet}(y)$ make it clear that the characteristic expansion parameter is not $C_F \alpha_s/\pi$ but more like $C_F(\alpha_s/\pi)\ln^2 y$. For $C_F(\alpha_s/\pi)\ln^2 y < 1$ we must have $y > 0.02$ (for $\alpha_s/\pi = 0.05$) which shows again that y must be large enough to have a small expansion parameter. If y is much smaller than this, finite order perturbation theory breaks down, and we must sum all powers of α_s/π. This is impossible in general, but it can be done if we restrict ourselves to the leading terms $\sim(\alpha_s/\pi)\ln^2 y$. Then we get for the sum

$$\sigma^{2-jet}(y) = \sigma^{(2)} \exp\left(-\frac{\alpha_s}{\pi} C_F \ln^2 y\right) .$$

(3.1.77)

and for the 3-jet cross section

$$\sigma^{3-jet}(y) = \sigma^{(2)} \frac{\alpha_s}{\pi} C_F \ln^2 y \cdot \exp\left(-\frac{\alpha_s}{\pi} C_F \ln^2 y\right) .$$

(3.1.78)

These formulae go over into (3.1.68) and (3.1.69) if we expand up to $O(\alpha_s)$. We see that $\sigma^{2-jet}(y)$ is positive, independent how small y is, and that it vanishes for $y \to 0$. The same is true for $\sigma^{3-jet}(y)$. This means that for a vanishing cut-off value y all exclusive multi-jet cross sections vanish and only the cross section with an infinite number of jets is unequal zero, a result quite familiar from QED.

Unfortunately the approximation of $\sigma^{2-jet}(y)$ and $\sigma^{3-jet}(y)$ by the leading logarithmic terms in lny is a very bad approximation for $y \simeq 0.05$ and can be valid only for very small y. So it cannot be used in practice. The approximations (3.1.77) and (3.1.78) as a function of y^{-1} are also shown in Fig.3.14 [we divided (3.1.77) and

(3.1.78) by $\sigma^{(2)}$]. Of course, these curves give only a qualitative insight how σ^{2-jet} and σ^{3-jet} might behave for smaller y's. We see that $\sigma^{3-jet}(y)$ as given by (3.1.78) remains smaller than 0.4 $\sigma^{(2)}$ for all y, which we may take as a further hint, that in $O(\alpha_s)$ y should be chosen in such a way that $\sigma^{3-jet}(y)$ is only a fraction of $\sigma^{(2)}$.

3.1.8 The Scalar Gluon Model

Although there is no other field theory in sight for describing the interaction of quarks and gluons which has the same attractive features as QCD, i.e. the underlying gauge structure and asymptotic freedom, it is of interest to study whether other gluon theories could explain the experimental data on jets as well as QCD. It is clear that on the level of lowest order perturbation theory the non-abelian SU(3) colour theory is not the only candidate. At this level the non-abelian structure does not come in and the abelian vector gluon theory leads to identical predictions. Therefore, in order to have a second candidate, the scalar gluon theory was considered, which has different predictions already in $O(\alpha_s)$ of perturbation theory.

The scalar gluon model is defined by the interaction Lagrangian

$$\mathscr{L}_I = g^* \bar{q} \lambda_a \phi_a q \tag{3.1.79}$$

where ϕ_a describes the scalar gluon field with colour quantum number a. For zero-mass quarks all results of scalar gluons are identical to those of pseudoscalar gluons because of γ_5 invariance.

The cross section $d^2\sigma/dx_1 dx_2$ for $e^+ e^- \to q\bar{q}g$ with a scalar gluon g was calculated by Ellis, Gaillard and Ross /1976,1977/ and has the following form

$$\frac{d^2\sigma}{dx_1 dx_2} = \sigma^{(2)} \frac{\alpha_s^*}{4\pi} C_F \frac{x_3^2}{(1-x_1)(1-x_2)} \tag{3.1.80}$$

where $\alpha_s^* = g^{*2}/4\pi$ and $\sigma^{(2)}$ is given in (3.1.12). This has to be compared to (3.1.11) for vector gluons. The right hand side of (3.1.80) is less singular for $x_1, x_2 \to 1$. But this region is occupied by the 2-jet events. In the genuine 3-jet region $x_1 \simeq x_2 \simeq 2/3$ the difference between (3.1.11) and (3.1.80) is much less apparent.

The other partial cross sections are /Kramer, 1980; Schierholz, 1979; Hoyer, Osland, Sander, Walsh and Zerwas, 1979/; for $m_q \neq 0$ see Laermann and Zerwas /1980/ using the notation of Sect.3.1.2:

(i) $\mathbf{p_1} \parallel \overrightarrow{0z}$

$$\frac{d^2\sigma_L}{dx_1 dx_2} = \sigma^{(2)} \frac{\alpha_s^*}{4\pi} C_F \frac{2(1-x_3)}{x_1^2}$$

$$\frac{d^2\sigma_T}{dx_1 dx_2} = \frac{1}{2} \frac{d^2\sigma_L}{dx_1 dx_2}$$

$$\frac{d^2\sigma_I}{dx_1 dx_2} = \sigma^{(2)} \frac{\alpha_s^*}{4\pi} C_F \left[\frac{1-x_3}{2(1-x_1)(1-x_2)} \right]^{1/2} \left[1 + \frac{x_1 x_2 - 2(1-x_3)}{x_1^2} \right] \qquad (3.1.81)$$

(ii) $\mathbf{p_2} \parallel \overrightarrow{0z}$ = (i) with $x_1 \leftrightarrow x_2$ \qquad (3.1.82)

(iii) $\mathbf{p_3} \parallel \overrightarrow{0z}$

$$\frac{d^2\sigma_L}{dx_1 dx_2} = \frac{d^2\sigma_T}{dx_1 dx_2} = \frac{d^2\sigma_I}{dx_1 dx_2} = 0 \quad . \qquad (3.1.83)$$

Since the other cross sections vanish for the case that the gluon momentum is chosen as the z axis in the jet plane, one has a very clean way of testing the spin of the gluons provided the gluon jet could be identified. For scalar gluons the hadron distribution around the gluon jet axis is isotropic. For vector gluons we expect a non-isotropic azimuthal distribution around the gluon jet axis which follows from (3.1.15) (see Kramer, Schierholz and Willrodt /1978,1979/ for further details). It is clear that the more complicated θ and χ dependence based on (i) and (ii) is due to the particular choice of quark and antiquark momentum as reference axis, whereas with the gluon as reference axis the angular distribution is of the form $(1 + \cos^2\theta)$ as for $e^+e^- \rightarrow q\bar{q}$.

Similar to the vector gluon case various single variable distribution have been derived from (3.1.80) to (3.1.83). The thrust distribution $d\sigma/dT$ obtained after integrating over θ and χ is /De Rujula, Ellis, Floratos and Gaillard, 1978/

$$\frac{1}{\sigma^{(2)}} \frac{d\sigma}{dT} = \frac{\alpha_s^*}{4} C_F \left[2\ln(\frac{2T-1}{1-T}) + \frac{(4-3T)(3T-2)}{1-T} \right] \quad . \qquad (3.1.84)$$

This has to be compared to (3.1.18) for the vector gluon case. (3.1.84) is less singular for $T \rightarrow 1$ than (3.1.18).

The formula for the average $(1-T)$ deduced from (3.1.84) is

$$<1-T> = \frac{\alpha_s^*}{4\pi} C_F \left(\frac{1}{18} + \frac{1}{4} \ln 3\right) = 0.110 \frac{\alpha_s^*}{\pi} \quad . \tag{3.1.85}$$

The cross sections analogous to (3.1.22) are:

$$\frac{1}{\sigma^{(2)}} \frac{d\sigma_L}{dT} = 2 \frac{1}{\sigma^{(2)}} \frac{d\sigma_T}{dT} = \frac{\alpha_s^*}{4\pi} C_F \frac{2(3T-2)}{T}$$

$$\frac{1}{\sigma^{(2)}} \frac{d\sigma_I}{dT} = \frac{\alpha_s^*}{4\pi} C_F \sqrt{2} \left[\frac{2(1-T)\sqrt{2T-1}}{T} - \sqrt{1-T}\right] \quad . \tag{3.1.86}$$

They follow from (3.1.81-83) with the convention that the hadron plane is defined by the thrust axis and the momentum of the second most energetic jet as in (3.1.22) /Kramer, 1980; Schierholz, 1979/. Plots of these cross sections and comparisons with the vector gluon theory can be found in the work of Hoyer, Osland, Sander, Walsh and Zerwas /1979/. If the Yukawa coupling α_s^* is adjusted so that $d\sigma_U/dT$ (scalar) = $d\sigma_U/dT$ (vector) at T = 0.8 one finds that $d\sigma_L/dT$ is larger in the scalar gluon case than in the vector gluon case whereas $d\sigma_I/dT$ is smaller.

Another way to compare the two theories is based on the coefficients $\alpha(T)$, $\beta(T)$ and $\gamma(T)$ in the angular distribution $w(\theta,\chi)$ defined in (3.1.24). They are plotted as a function of T, for the two cases, vector and scalar gluon in Fig.3.6.

The $d\sigma/dx_\perp$ distribution is obtained from (3.1.29) where $d^2\sigma/dTdx_\perp^2$ is:

$$\frac{1}{\sigma^{(2)}} \frac{d^2\sigma}{dTdx_\perp^2} = \frac{\alpha_s^*}{4\pi} C_F \frac{T}{4(1-T)[1-x_\perp^2/(1-T)]^{1/2}}$$

$$\left\{2\left[\frac{(2-T-x_{2+})^2}{(1-T)(1-x_{2+})} + \frac{(2-T-x_{2-})^2}{(1-T)(1-x_{2-})}\right] + \frac{T^2}{(T+x_{2+}-1)(1-x_{2+})} + \frac{T^2}{(T+x_{2-}-1)(1-x_{2-})}\right\} \tag{3.1.87}$$

and where $x_{2\pm}$ is defined in (3.1.33). The moment $<x_\perp^2>$ is in lowest order

$$<x_\perp^2> = \frac{\alpha_s^*}{\pi} C_F \left(\frac{59}{9} - 16\ln\frac{3}{2}\right) = 0.0908 \frac{\alpha_s^*}{\pi} \tag{3.1.88}$$

so that

$$\frac{<1-T>}{<x_\perp^2>} = \begin{cases} 1.21 & \text{for scalar gluons} \\ 1.57 & \text{for vector gluons} \end{cases}$$

independent of the value of α_s^* or α_s. But as will be discussed later these moments are very much influenced by hadronization effects so that their perturbative measures are a bad criterion for the underlying theory.

Another variable which played some role in distinguishing vector and scalar theories in lowest order is the following variable $\sin\tilde{\theta}$

$$\sin\tilde{\theta} = \frac{x_\perp}{\sqrt{1-T}} \qquad (3.1.89)$$

first introduced by Ellis and Karliner /1979/. For $q\bar{q}g$ events this variable is bounded by

$$\frac{4(1-T)(2T-1)}{T^2} \leq \sin^2\tilde{\theta} \leq 1 \qquad (3.1.90)$$

which follows from (3.1.30). As an example, if $T = 0.8$ the bound (3.1.90) forces $\tilde{\theta}$ to lie between 60° and 90°. Actually $\tilde{\theta}$ has a nice kinematic interpretation. It is the angle formed between the less energetic jet and the thrust axis in the Lorentz system where the two less energetic jets emerge in their center-of-mass frame back-to-back.

Their total energy is $E = W(1-T/2)$ and their invariant mass is $M = W\sqrt{1-T}$ so that the Lorentz boost is obtained from

$$\tanh\zeta = \frac{T}{2-T} \quad . \qquad (3.1.91)$$

The angular distribution in $\tilde{\theta}$ has been calculated by Ellis and Karliner /1979/ from the cross sections (3.1.11) and (3.1.80). They normalized it to 1 at $\cos\tilde{\theta} = 0$. Then for scalar gluons

$$N\frac{d\sigma}{d\cos\tilde{\theta}} = \frac{4-3T^2+T(3T-4)\cos^2\tilde{\theta}}{(4-3T^2)(1-\cos^2\tilde{\theta})} \qquad (3.1.92)$$

and for vector gluons

$$N\frac{d\sigma}{d\cos\tilde{\theta}} = \frac{4T^3+(2-T)^3+3(2-T)T^2\cos^2\tilde{\theta}}{[4T^3+(2-T)^3](1-\cos^2\tilde{\theta})} \quad . \qquad (3.1.93)$$

The distribution (3.1.93) for vector gluons is essentially independent of T, with the exception of the change in the kinematic bound (3.1.90). In the scalar case the angular distributions depend slightly on T becoming flatter as T increases.

Unfortunately these distributions cannot be compared directly to experimental data. The latter refer to hadron distributions whereas (3.1.92) and (3.1.93) give the angular distribution for parton directions. We shall see in Sect.3.1.9 how the distributions are modified when fragmentation effects are taken into account.

3.1.9 Evidence for $q\bar{q}g$ Production, Comparison with Experimental Data

In this section we shall present some examples for a comparison of experimental data with $O(\alpha_s^2)$ perturbation theory results, in particular with the formula $d\sigma^2/dx_1 dx_2$ in Sect.3.1.2. Many of such comparisons have been reported since 1979 when the evidence for $q\bar{q}g$ arose the first time. This is documented in the papers: Bartel et al. /1980/, Barber et al. /1979/, Berger et al. /1979/, Brandelik et al. /1979/. With increasing statistics this evidence became more and more convincing as time went on. The later comparisons are found in various original articles, conference proceedings and review papers. Besides those referred to already in the preface we mention a few more: Wolf /1980,1981/, Söding /1981/, Duinker /1982/, Criegee and Knies /1982/, The MARK J Collaboration /1980/, Wu /1981/ and Saxon /1982/.

When considering such comparisons in more detail we shall discuss in particular the problems caused by the fact that experimentally not quarks and gluons but hadrons, pions, kaons, protons, etas etc. represent the final state. To a large extent the hadrons are collimated into jets. But even at the highest PETRA energies these jets are still rather broad. This broadening of the jets caused by the fragmentation of quarks and gluons into hadrons due to a finite transverse momentum leads, for example, to an appreciable change of the thrust distribution (3.1.18) predicted by perturbation theory so that a direct confrontation of perturbation theory results with measured hadron distributions is not possible. The thrust values computed from the momenta of the produced hadrons in an event differ appreciably from the thrust values computed from the momenta of the 3 jets, q, $\bar{q}$ and g which produced the event. In addition a large fraction of the measured events are 2-jet events which due to fragmentation effects and further broadening due to weak decays look like 3-jet events. It is clear that a direct confrontation of experimental data with theory is possible only if all these effects, as $q\bar{q}$ background and non-perturbative broadening of jets can be described reliably with phenomenological models. Then starting from the $q\bar{q}g$ 3-jet formula and applying these models all hadron distributions are computed and compared with data. This road was taken after the first evidence for a third jet appeared. Before we discuss this approach in more detail we shall first consider the more direct way which was developed somewhat later.

The idea is to obtain information about jet multiplicities, jet distributions, jet-jet correlations etc. from the measured hadron events, which then can be directly compared with the predictions of QCD perturbation theory. In order to achieve this one needs effective procedures for identifying and reconstructing jets in statistically large samples of events. These methods are usually referred to as cluster algorithms and have been developed by several authors: Lanius /1980/, Dorfan /1981, Daum, Meyer and Bürger /1981/, Lanius, Roloff and Schiller /1981/, Goddard /1981/,

Bäcker /1981/, Yamamoto /1981/, Babcock and Cutkosky /1981,1982a,1982b/. The methods described in these papers differ somewhat concerning the criteria how jets are selected. The first application to e^+e^- data was done by the PLUTO Collaboration at PETRA /Berger et al., 1980a/ based on the algorithm of Daum, Meyer and Bürger, which we shall describe now.

In this method jets are selected on the basis of angle criteria. In a first step all particles in an event (charged and neutrals, in case neutral particles have been measured) with momenta, which are less than 30° apart, are combined into preclusters. Single isolated particles are also considered as preclusters. The sum of the particle momenta in a precluster defines the momentum of the precluster. In a second step all preclusters whose momenta have an angle less than δ are combined to clusters. Single isolated preclusters are considered as clusters. Then all clusters with a total energy more than 2 GeV and with at least two particles are called jets if at least (1-ε/2) of the total energy is contained in the clusters. The number of jets in an event found this way gives the jet multiplicities and the sum of the particle momenta in a jet gives the momentum **p** of the jet. The angle δ which was introduced for combining preclusters is correlated with the angle δ introduced in connection with the Sterman-Weinberg definition of jets in Sect.3.1.7. In the same way the parameter ε used in the cluster method to define jets corresponds to the Sterman-Weinberg parameter ε. Analogous to the cluster algorithm based on angle criteria the JADE Collaboration at PETRA /Bartel et al. 1982a/ has developed the cluster method based on the invariant mass of two particles. Instead of the angle between the two particle momenta $\mathbf{p}_i$ and $\mathbf{p}_j$ they determined the invariant mass squared $y_{ij} = (p_i+p_j)^2/W^2$. If y_{kl} is the smallest of all y_{ij} in an event, the particles k and l are combined to a pseudo particle with four-momentum $(|\mathbf{p}_k|+|\mathbf{p}_l|,\mathbf{p}_k+\mathbf{p}_l)$. This procedure is repeated until y_{kl} is larger than a limit value y. The number of pseudo particles which remain at this step is the number of jets in an event and their momenta are equal to the jet momenta.

It is clear that the distribution of jet multiplicities obtained with the cluster algorithm depends on the chosen parameters ε, δ or y, respectively. For small ε, δ or y one necessarily finds very many jets, which mostly come from fluctuations due to the fragmentation of quarks and gluons into hadrons. For large ε, δ or y, on the other hand, we find essentially only 2-jet events whereas 3-jets (or more than 3) are not resolved. Detailed studies using the phenomenological fragmentation models described in Sect.2.4.2 have revealed the range of ε, δ or y values, for which jets originating from $q\bar{q}g$ intermediate states are resolved and the pure fragmentation effects are suppressed. Such values are ε = 0.2, δ = 40° and y = 0.04. For example, with y = 0.04 one finds experimentally less than 4% of the events have more than 3 jets, roughly 30% of the events have 3 jets and nearly 70% are 2 jets. The limitation of the y_{k1} by y corresponds also to the parameter y introduced in connection with the separation of the $q\bar{q}g$ phase space in a 2-jet and a 3-jet region. Here y was the boundary for combining q and g or $\bar{q}$ and g, respectively, to 2 jets. If we use (3.1.68) and (3.1.69) to calculate the perturbative jet multiplicities we obtain for $y = 0.04$ and $\alpha_s/\pi = 0.05$ 66% as 2-jet rate and 34% for the 3-jet rate in good agreement with the result found experimentally.

We show in Fig.3.15 the first results of a cluster analysis with the angle method. The data come from the PLUTO Collaboration at PETRA /Berger et al., 1980a/ and are compared with the theoretical curve (3.1.18) ("vector gluon" in the figure). The $q\bar{q}g$ thrust T is denoted by x_1, the maximal jet energy. The agreement between the experimental points and the theoretical curve is very good, but less with a curve labelled "scalar gluon". This curve is obtained from (3.1.84) in Sect.3.1.8 where further details about the scalar gluon model are found. The QCD coupling constant obtained from these data is $\alpha_s = 0.15$. A similar comparison with more recent TASSO data can be seen in Fig.3.16. With these data, which have much better statistical accuracy, the scalar gluon model can be definitely excluded. Also another toy model, the so-called constituent interchange model (CIM), invented by De Grand, Ng and Tye /1977/ is in disagree-

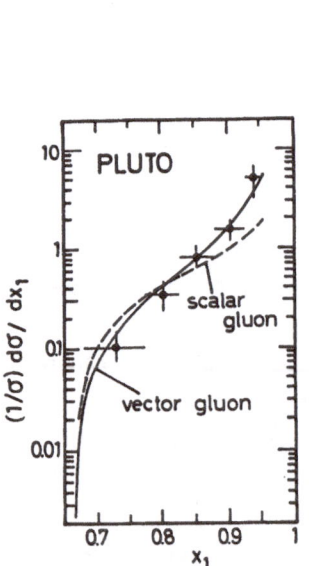
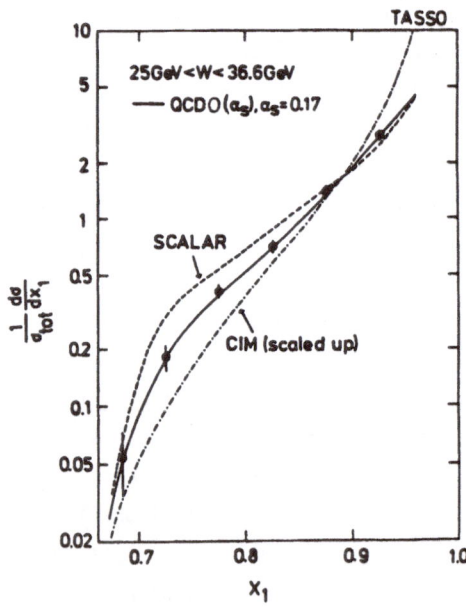

Fig.3.15. Thrust distribution $(1/\sigma)d\sigma/dx_1$ for W = 30 GeV obtained from cluster analysis of PLUTO data compared to vector and scalar gluon theory. $\alpha_s = 0.15$, $x_1 \equiv T$ is the thrust of $q\bar{q}g$ final state

Fig.3.16. Distribution of the maximum jet energy x_1 (= parton thrust T) at W = 33 GeV measured by TASSO compared to QCD ($\alpha_s = 0.17$, full curve), scalar gluon theory (dashed curve) and constituent interchange model (CIM, dashed-dotted curve)

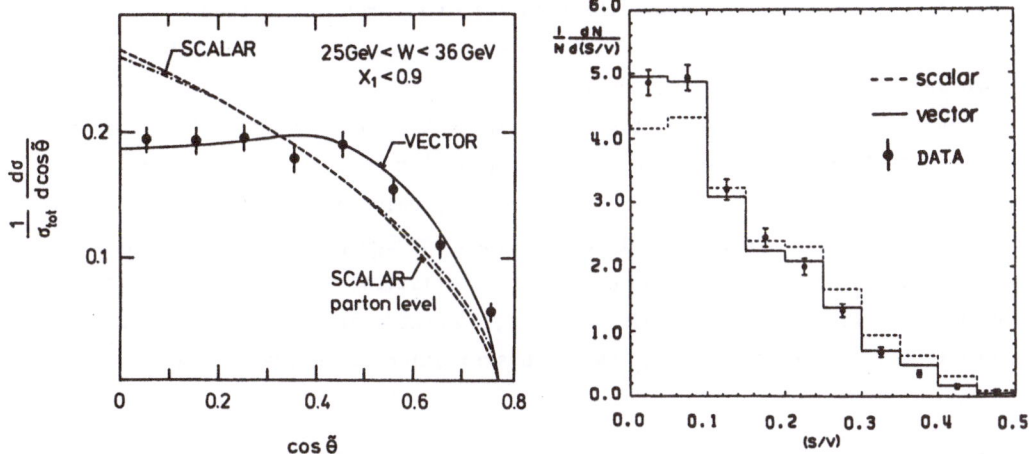

Fig.3.17. Observed distribution $(1/\sigma)d\sigma/d\cos\tilde\theta$ in the region $x_1 < 0.9$ as a function of the Ellis-Karliner angle $\tilde\theta$. The solid line shows QCD and fragmentation model prediction. The dashed line is the prediction for scalar gluon model with fragmentation, dashed-dotted line is for scalar model on the parton level

Fig.3.18. Observed distribution of the MARK J data as a function of the variable $x_3^2/(x_1^2+x_2^2)$. The solid line shows QCD fragmentation (Ali) model prediction. Dotted line is for scalar gluons

ment with the data. The QCD coupling constant adjusted to these data is $\alpha_s = 0.17$ /Wolf, 1982/. Here the variable is denoted $x_1 = T$, the maximal jet energy. Further tests on the gluon spin are shown in Figs.3.17 and 3.18. These tests are based on the high statistics available now and were performed by the TASSO /1982/ and the MARK J /1982/ Collaborations at PETRA. In Fig.3.17 the events are plotted as a function of the Ellis-Karliner angle $\tilde\theta$ which was introduced in Sect.3.1.8 and compared with the model predictions under the two assumptions of spin 1 and spin 0. Spin 0 is clearly ruled out. In this plot all events with $x_1 < 0.9$, where x_1 is the momentum of the most energetic jet, are included. The angle $\tilde\theta$ was determined from the jet energies x_i, arranged in the order $x_3 < x_2 < x_1$, so that $|\cos\tilde\theta| = (x_2-x_3)/x_1$. The MARK J Collaboration used the ratio $S/V = x_3^2/(x_1^2+x_2^2)$ instead of $|\cos\tilde\theta|$ to plot their results. S/V is just the ratio of cross sections for scalar and vector gluon production [see (3.1.11) and (3.1.80)]. Also with this method spin 0 for the gluon is ruled out. Similar conclusions reached on data with less statistics are reported in Brandelik et al. /1980b/ and Behrend et al. /1982a/.

The most elaborate cluster analysis was performed by the JADE Collaboration at PETRA /Bartel et al., 1982a/. They used both methods described above, the angle method and the method employing invariant masses. Furthermore, they produced data for two distributions, the $T = x_1$ distribution and the distribution in the variable $x_\perp$ introduced in Sect.3.1.4, i.e. $x_\perp = x_2\sin\theta_{12} = x_3\sin\theta_{13} = (2/x_1)[(1-x_1)(1-x_2)(1-x_3)]^{1/2}$, where x_1 is the maximal jet energy. The data were compared to the theoretical distribution (3.1.18) and to (3.1.29) which was already shown in Fig.3.8. The data will be shown in the next section in connection with a comparison with theory up to $O(\alpha_s^2)$. The energy is W = 33.8 GeV. The fit of the data to the thrust formula (3.1.18) and to the $x_\perp$ distribution (3.1.29) has given the coupling constants α_s in Table 3.1, referred to as α_s (first-order). For determining α_s only the data for $x_1 \leq 0.85$ and $x_\perp \geq 0.30$ were used in order to avoid distortions through possible 2-jet contributions at large x_1 and small $x_\perp$. The errors are statistical only. We see that the α_s values obtained with the two methods, y and ε, δ method, are in good agreement with each other. Also the fits to the x_1 and $x_\perp$ distributions give the same α_s.

Table 3.1. Strong coupling constant α_s in first order and first and second order QCD as obtained by the JADE Collaboration from cluster analysis

Jet Definition	Variable	α_s (first order)	α_s (first and second order)
Invariant Mass	x_1	0.208 ± 0.015	0.163 ± 0.010
	$x_\perp$	0.205 ± 0.013	0.158 ± 0.010
Angle δ	x_1	0.196 ± 0.013	0.170 ± 0.010
Energy ε	$x_\perp$	0.195 ± 0.013	0.175 ± 0.012

The systematical error in the α_s determination is 0.03. It has several causes. First the results depend somewhat on the definition of jets, i.e. how the angle δ is chosen to define preclusters etc. Another reason is the influence of fragmentation models. The results are not completely independent from these models as it may appear. The measured distributions must be corrected for the fact that not all 3-jet events come from $q\bar{q}g$ intermediate states but also from $q\bar{q}$ states including weak decays etc. The correction factors were obtained from Monte Carlo results based on the model of Ali et al. /1980/ and the Lund model. The final results depended rather weakly on the model used. This is a great advantage of the clusters algorithm. Therefore the results for α_s in Table 3.1 are rather independent of fragmentation models. This result, however, is not consistent with a result of the CELLO Collaboration at PETRA /Behrend et al., 1983/, that the coupling constant α_s obtained from fits to various distributions depends strongly on the fragmentation model. If one looks only at their result for the cluster method, the spread in α_s obtained with the two models is 0.08 ± 0.04, which is somewhat larger than the JADE spread in α_s. The origin of this difference is still unclear. It is conceivable, that with more and better experimental data available, one will be able to pin down the parameters of the models or even eliminate some of them (see for example Bartel et al. /1981, 1983/) so that the systematic error on α_s will diminish. But it should be clear that the cluster methods give us the most direct way to compare perturbative QCD results with empirical data. However, it is important to realize that in this method the momentum of the cluster is identified with the original parton momentum which need not be the case necessarily. It seems that in the Lund model a distortion of the cluster direction and the parton direction takes place, which produces the differences in the α_s values obtained.

As the next topic we shall discuss such comparisons of theory with experimental data which are based on directly measured hadron distributions. Theoretically these hadron distributions are computed with the help of the fragmentation models described in Sect.2.4.2 and with the $q\bar{q}g$ cross section as input. First we shall consider distributions in jet variables like thrust etc. For such thrust distributions, $(1/\sigma)d\sigma/dT$, the thrust T is determined according to the formula (2.3.4) from the measured momenta of charged and neutral particles (if neutral particles could be measured). How such an event looks in reality is shown in Fig.3.19. This is one of the first events which looks like a real 3-jet event. It was found with the TASSO detector. Shown are the momenta of charged particles projected into three mutually perpendicular planes /Brandelik et al. 1979/. In general, T varies between T = 1/2 (isotropic event) and 1 (collinear event). A thrust distribution of measured events is shown in Fig.3.20. These data which come from the JADE Collaboration /Elsen, 1981/ are corrected for acceptance losses and can be compared directly to theory. The $q\bar{q}g$ thrust distribution (3.1.18) with α_s = 0.18 is plotted also in this figure. There is no

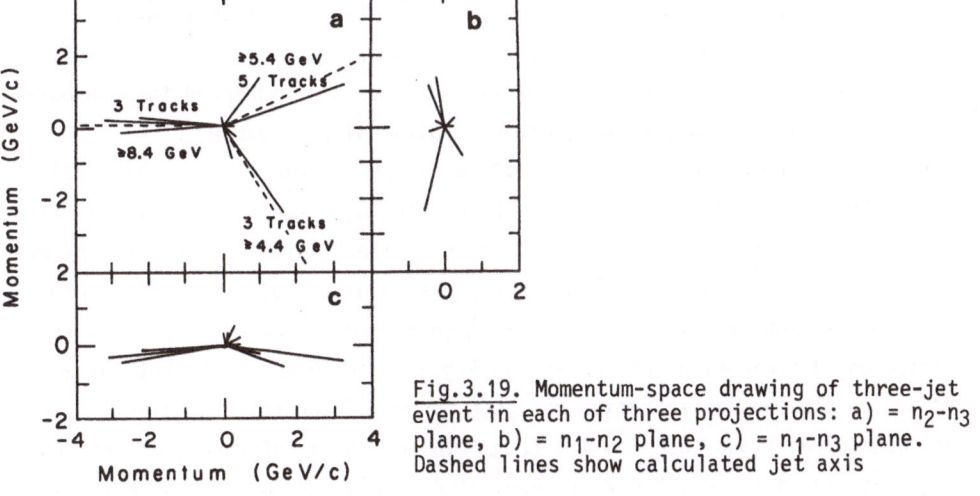

Fig.3.19. Momentum-space drawing of three-jet event in each of three projections: a) = n_2-n_3 plane, b) = n_1-n_2 plane, c) = n_1-n_3 plane. Dashed lines show calculated jet axis

Fig.3.20. Experimental data for thrust distribution of hadrons (JADE) compared to $q\bar{q}g$ thrust distribution with $\alpha_s = 0.18$

Fig.3.21. Observed thrust distributions of the JADE Collaboration for W = 30 and 35 GeV compared with various models. a) and c): $q\bar{q}$ model (dashed curve), $q\bar{q} + q\bar{q}g$ according to Hoyer model (full curve); b) and d): $q\bar{q} + q\bar{q}g$ according to Ali model (dashed curve) and according to Lund model (full curve)

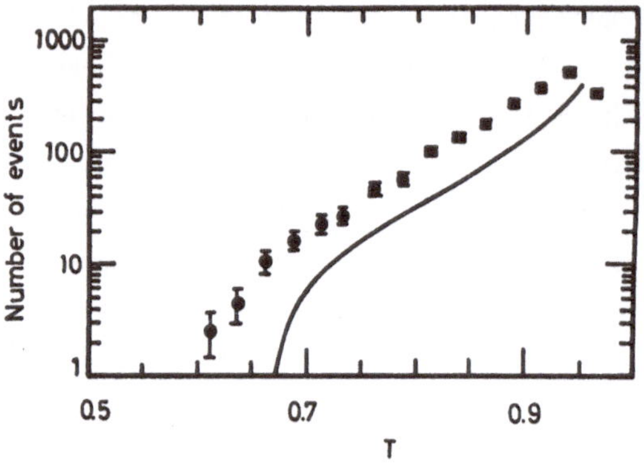

a)

b)

W = 30 GeV   W = 30 GeV

c)

d)

W = 35 GeV   W = 35 GeV

Fig.3.20

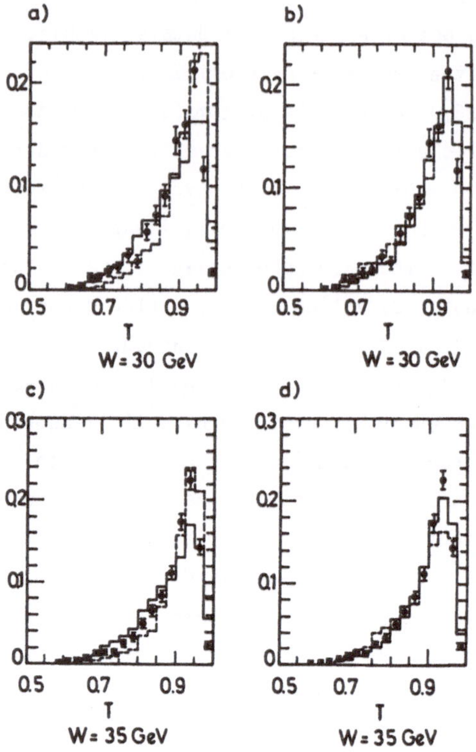

Fig.3.21

agreement between data and the theoretical distribution. For medium T values the data lie higher than a factor of 3, and there are also events in the region T < 2/3, which is kinematically forbidden for a 3-quantum final state. The reason for this disagreement, of course, is that the experimental thrust distribution has a large background originating from $q\bar{q}$ events, which lead to thrust distributions of finite width like in Fig.2.17. This broadening of the $q\bar{q}$ distribution which is due to the finite transverse momentum in the fragmentation of quarks is enhanced through mass effects, i.e. the finite b quark mass, and by additional weak decays of c and b quarks (or of the hadrons containing these quark flavours) /De Rujula et al., 1978; Ali et al., 1979a,1979b,1980/. In addition, when the partons q, $\bar{q}$ and g fragment into hadrons the resulting T distribution is shifted to smaller T's as compared to the perturbative curve in Fig.3.20. How the non-perturbative distribution resulting alone from the fragmentation of q and $\bar{q}$ in $e^+e^+ \rightarrow q\bar{q}$ (q = u, d, s, c, b) compares with the perturbative $q\bar{q}g$ distribution can be seen in Fig.3.4. These two distributions are then added after the $q\bar{q}g$ distribution has been corrected for fragmentation. The results of such calculations based on the fragmentation models of Hoyer, Ali and Lund which were introduced in Sect.2.4.2 are exhibited in Fig.3.21 and compared to experimental data of the JADE Collaboration /Elsen, 1981/. Also shown is the distribution for a pure $q\bar{q}$ model with five quarks (Fig.3.21a,c). It is clearly seen that the data lie well above the $q\bar{q}$ curves for T < 0.9. Concerning the fragmentation models for $q\bar{q}$ + $q\bar{q}g$ all three models describe very well the enhancement for T < 0.85. It seems that the Lund model produces the best overall fit of the data for W = 30 and 35 GeV.

When computing the distribution of thrust or of any other jet variable for the $q\bar{q}g$ component one must remember the problem of 2- and 3-jet separation. The starting point is the formula for the $q\bar{q}g$ cross section (3.1.11) from which the distribution of parton momenta x_1 and x_2 follows. In Sect.3.1.7 we discussed already that in the kinematic region $1-y \leq x_1$, $x_2 \leq 1$ the cross section is part of the 2-jet cross section. The 2-jet cross sections is, however, covered already by the non-perturbative $q\bar{q}$ contribution. Therefore in the $q\bar{q}g$ contribution the region $1-y = T_0 \leq x_1$, $x_2 \leq 1$ is discarded. This way one avoids double counting of the 2-jet contribution and also the singular region of (3.1.11) at $x_1, x_2 = 1$. The cut-off parameter chosen is $T_0 = 0.95$ in accordance with the discussion in Sect.3.1.7. The normalization of the two components follows from the fact that their sum must be equal to $\sigma_{tot} = \sigma^{(2)}(1+\alpha_s/\pi)$. Clearly the degree of broadening of the T distribution caused by the addition of the $q\bar{q}g$ part depends on the value of α_s. The larger α_s is the more the T distribution is shifted to smaller T's. Fitting the model with both components $q\bar{q}$ and $q\bar{q}g$ then allows to determine α_s.

It is obvious that the broadening effect caused by the $q\bar{q}g$ component increases with increasing energy W. With increasing W the non-perturbative $q\bar{q}$ contribution becomes narrower and narrower (see Fig.2.17) in accord with $<1-T>_{non-pert.} \sim 1/W$, whereas the $q\bar{q}g$ contribution is practically energy independent since it is proportional to $\alpha_s \sim 1/\ln(W^2/\Lambda^2)$. Therefore, for energies below 25 GeV, there is practically no sign for a third jet, the gluon jet, since it is hidden completely in the non-perturbative jet spread of the $q\bar{q}$ component. How thrust distributions becomes narrow with increasing c.m. energy is seen in Fig.3.22b. Shown are older data of the TASSO Collaboration /Brandelik et al., 1980c/ for W's between 13 and 31.2 GeV. At 13 GeV the distribution is rather broad with the maximum at T = 0.8. Above 27.6 GeV the distributions become narrow. The same can be said about the sphericity distributions shown in Fig.3.22a for the same energies. The curves are QCD predictions based on the Hoyer model with five or six quarks. This analysis had the purpose (in 1979) to yield a signal for the production of Q = 2/3 top quarks, which are seen not to be present due to the disagreement of the data with the 6 quark curves. The results of a similar study for the highest PETRA energies reached at the end of 1982 with W's in the interval $37.94 \leq W \leq 38.63$ is reproduced in Fig. 3.23. The data come from the MARK-J Collaboration at PETRA /Adeva et al., 1983b/.

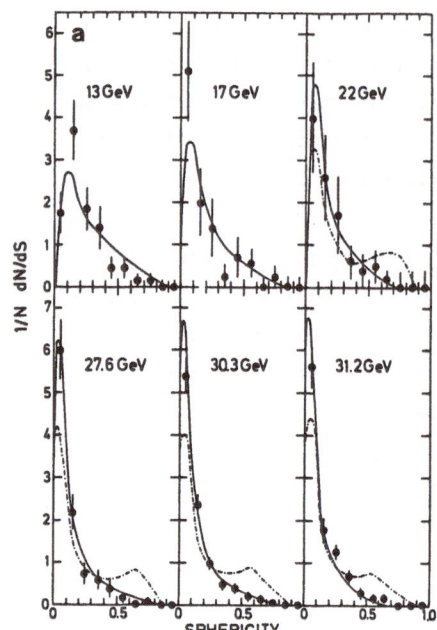
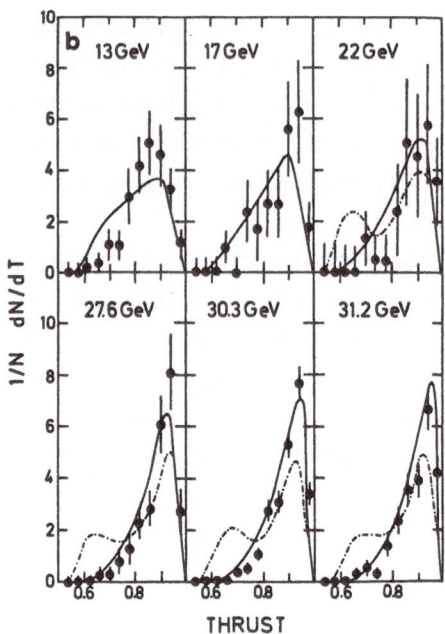

Fig.3.22. a) Observed (TASSO) spericity distribution for various energies compared to Hoyer model with 5 and 6 (dashed-dotted curve) quarks. b) Same as a) for thrust

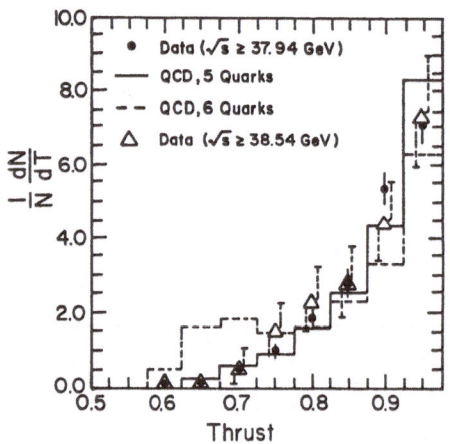

Fig.3.23. Observed thrust distribution (MARK J) for highest PETRA energy compared to Ali model with 5 and 6 quarks

The measured thrust distributions agree well with the distributions at lower energies and with Monte-Carlo computations based on the Ali model with 5 quarks and also do not show any sign for production of open top quarks.

The broadening of the thrust distribution due to the emission of a third jet is, as we have seen in Fig.3.21, not very strong. The main reason for this is the 2-jet contribution from $q\bar{q}$ which, even above W = 30 GeV, is still very broad. There are other variables, where the $q\bar{q}$ contribution is narrower in relation to the $q\bar{q}g$ part. Such a variable is m_h^2/W^2, where m_h is the invariant mass of the heavy jet. To determine m_h, a jet axis is computed for each event, for example, on the basis of thrust. Then a plane perpendicular to the thrust axis is chosen. The invariant masses on both sides of this plane are ordered and give m_h and m_1, m_1 is the mass of the light jet. For $q\bar{q}g$ final states we have $m_h^2 = W^2(1-T)$ and $m_1^2 = 0$. Thus m_1^2 is a measure of the non-perturbative width of the jet whereas m_h^2 contains the effect of the $q\bar{q}g$ contribution and additional non-perturbative effects. It was found by Elsen /1981/ that the background in the m_h^2 distribution coming from $q\bar{q}$ decreases much stronger with increasing m_h^2 than in the thrust distribution.

So far, only distributions in one jet variable have been analysed and compared to theory. It is clear that the influence of the $q\bar{q}g$ component could be enhanced by looking at distributions in two variables. In this case the comparison with theory could be limited to kinematical regions, where the $q\bar{q}g$ component is dominant. This is the case for x_1,x_2 values in the vicinity of 2/3. Since the cross section is very low there, the statistics is very limited in this region which is obvious also from the data in Figs.3.21-22.

Up to now we have considered only distributions in jet measures like thrust, sphericity etc. These variables describe essentially the global features of the momentum flow of all hadrons produced in e^+e^- annihilation. Another possibility is to trigger on specific hadrons, pions, K-mesons, protons and to study their momentum distributions. They also should be influenced by the existence of a third jet. One such distribution is the inclusive cross section $d\sigma/dx$ for the production of single hadrons as a function of the scaled momentum $x = 2p/W$ which was introduced in Sect.2.1. This cross section is, in first approximation [see (2.1.4)] given by the sum of fragmentation functions of quarks into single hadrons. Therefore it depends mostly on the fragmentation process and is less influenced by the emission of the gluon. Here the gluon emission produces the well-known scaling violations of the fragmentation functions referred to in the introduction. To see the effect of the gluon bremsstrahlung more pronounced, it is more profitable to study such quantities which are directly proportional to α_s. Examples are the angular distribution of singly produced hadrons with respect to the beam axis and the hadron transverse momentum with respect to a jet axis. In the following we shall give a simplified exposition of these two effects.

If we neglect the transverse momentum when a quark fragments into a hadron, then the hadron has the momentum direction of the quark. In this approximation the angular distributions of hadrons produced in the $e^+e^- \rightarrow q\bar{q}$ process is of the form of the jet axis distribution which is $(1+\cos^2\theta)$ as was discussed in Sect.2.1. In general, the cross section for single hadron production as a function of x and of θ, the angle between hadron momentum and beam axis, must be of the form (3.1.9):

$$\frac{d^2\sigma}{dx\,d\cos\theta} = \frac{3}{8}\,(1+\cos^2\theta)\,\frac{d\sigma_U}{dx} + \frac{3}{4}\,\sin^2\theta\,\frac{d\sigma_L}{dx} \tag{3.1.94}$$

$d\sigma_U/dx$ has a large component from $e^+e^- \to q\bar{q}$. But $d\sigma_L/dx$ can come only from $e^+e^- \to q\bar{q}g$ and subsequent fragmentation. This contribution can be calculated from the following expression /Politzer, 1977; Kramer and Schierholz, 1979/

$$\frac{d\sigma_L}{dx} = \int\limits_x^1 \frac{dx_1}{x_1}\,\frac{d\sigma_L}{dx_1}\,D_q\!\left(\frac{x}{x_1}\right) + \int\limits_x^1 \frac{dx_2}{x_2}\,\frac{d\sigma_L}{dx_2}\,D_{\bar{q}}\!\left(\frac{x}{x_2}\right) + \int\limits_x^1 \frac{dx_3}{x_3}\,\frac{d\sigma_L}{dx_3}\,D_g\!\left(\frac{x}{x_3}\right) \tag{3.1.95}$$

if the transverse momentum in the fragmentation of quarks and gluons can be neglected. $D_q(x)$, $D_{\bar{q}}(x)$ and $D_g(x)$ are the fragmentation functions of quark, antiquark and gluon into a hadron of momentum x. $D_q(x) = D_{\bar{q}}(x)$ can be determined in a first approximation by measuring $d\sigma/dx$ [see (2.1.4)]. The coefficients $d\sigma_L/dx_i$ (i = 1, 2, 3) in the integrands are derived from the longitudinal parts of the $q\bar{q}g$ cross sections in (3.1.13-15):

$$\frac{d\sigma_L}{dx_1} = \frac{d\sigma_L}{dx_2} = \frac{\alpha_s}{2\pi}\,C_F\,\sigma^{(2)}, \quad \frac{d\sigma_L}{dx_3} = \frac{\alpha_s}{2\pi}\,C_F\,\sigma^{(2)}\,\frac{4(1-x_3)}{x_3} \quad . \tag{3.1.96}$$

From (3.1.95) we derive the parameter $\alpha(x)$ in the single particle angular distribution $1 + \alpha(x)\cos^2\theta$

$$\alpha(x) = 1 - 4\,\frac{d\sigma_L/dx}{d\sigma/dx + d\sigma_L/dx} \quad . \tag{3.1.97}$$

Thus the difference $1-\alpha(x)$ is directly proportional to α_s and constitutes a direct measure of the $q\bar{q}g$ contribution [the denominator in (3.1.97) is $O(\alpha_s^0)$].

We have plotted $\alpha(x)$ in Fig.3.24 for fragmentation functions

$$D_q^c(x) = D_{\bar{q}}^c(x) = 2\,\frac{(1-x)^2}{x} \tag{3.1.98}$$

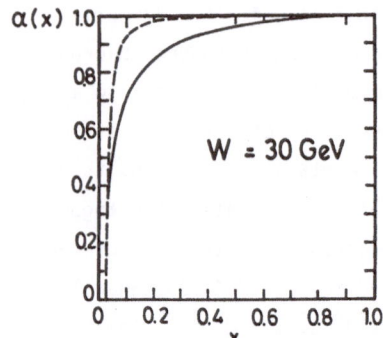

$W = 30$ GeV

Fig.3.24. The coefficient $\alpha(x)$ of the angular distribution $1+\alpha(x)\cos^2\theta$ for single hadrons calculated from $e^+e^- \to q\bar{q}g$ compared to non-perturbative contribution (dashed curve) with $\langle p_T^2\rangle = (0.3$ GeV$)^2$

and

$$D_g^c(x) = \int_x^1 \frac{dx_q}{x_q} D_q^c\left(\frac{x}{x_q}\right) + \int_x^1 \frac{dx_{\bar{q}}}{x_q} D_{\bar{q}}^c\left(\frac{x}{x_{\bar{q}}}\right) = 4\frac{1-x^2}{x} + 8\ln x \qquad (3.1.99)$$

where c stands for the sum over all charged particles. The gluon fragmentation function corresponds to the case where the gluon fragments first into a quark and an antiquark with a constant momentum distribution which then decays with fragmentation function (3.1.98). Also shown is the non-perturbative background which corresponds to the approximate formula

$$\frac{d\sigma_L}{dx} = 2\frac{<p_T^2>\text{non-pert.}}{x^2 W^2}\frac{d\sigma}{dx} \quad \text{or} \quad \alpha(x) = 1 - \frac{8<p_T^2>\text{non-pert.}}{x^2 W^2} \quad . \qquad (3.1.100)$$

$<p_T^2>\text{non-pert.}$ is the average transverse momentum of the fragmentation, which was chosen $(0.3\ \text{GeV})^2$. For smaller x we find a sizable (and lasting as W^2 increases) deviation of $\alpha(x)$ from 1 which is much bigger than what we expect from the non-perturbative background. So the measurement of $1-\alpha(x)$ for $x < 0.3$ is a signal for $q\bar{q}g$ effects. First experimental data from Brandelik et al. /1982/ are shown in Fig.3.25 compared with the theoretical curve of Fig.3.24. The data are consistent with theory but the statistical accuracy must be improved before further conclusions can be drawn. It is clear that $\alpha(x)$ could be measured with higher accuracy if transverse polarized beams were available [see (3.1.51)].

With similar simplifying assumptions about the fragmentation of quarks and gluons we can study the transverse momentum p_T in jets. Experimentally the transverse momenta of all hadrons in an event are measured with respect to the jet axis, which may be the thrust or the sphericity axis. For $q\bar{q}g$ final states the thrust axis is parallel

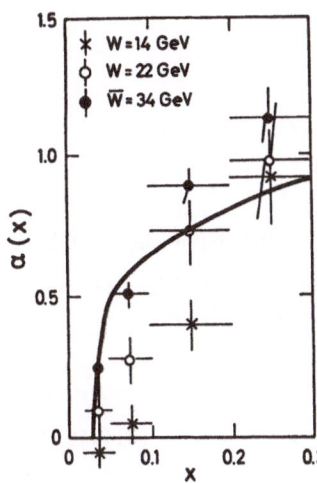

Fig.3.25. Experimental data for $\alpha(x)$ from TASSO for various energies compared to prediction from Fig.3.24

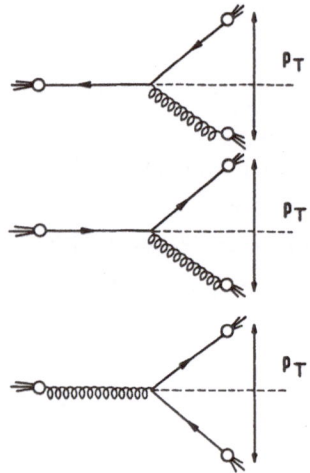

Fig.3.26. Kinematics of jet broadening caused by gluon emission. Dashed line represents jet axis

to the direction of the most energetic parton, which makes it the canonical axis for studying p_T phenomena. Then the hadron transverse momentum with respect to the thrust axis is given by the transverse momentum of the recoiling partons smeared with the fragmentation functions (see Fig.3.26) /Schierholz, 1979; Hoyer, Osland, Sander, Walsh and Zerwas, 1979; Ellis, Gaillard and Ross, 1976,1977; DeGrand, Ng and Tye, 1977/:

$$\frac{1}{\sigma}\frac{d^2\sigma}{dxdp_T^2} = \frac{\alpha_s}{8\pi}C_F\frac{1}{p_T^2}\frac{1-K^2}{K^2}\int_X^{\frac{2-\sqrt{2(1-K)}}{1+K}} dz \frac{1}{1-z} A^2 \left\{\left[\frac{A^2+z^2}{B(1-z)} + \frac{(1-z+B)^2+z^2}{(z-B)(1-z)}\right]\right.$$

$$\left. \cdot \left[D_q(\frac{x}{z}) + D_{\bar{q}}(\frac{x}{z})\right] + 2\frac{A^2+(1-z+B)^2}{B(z-B)} D_g(\frac{x}{z})\right\} \qquad (3.1.101)$$

where

$$K = (1-x_T^2/x^2)^{1/2} \quad , \qquad x_T = 2p_T/W$$

$$A = \frac{2(1-z)}{2-z(1+K)} \quad , \qquad B = \frac{z(1-K)}{2-z(1+K)}$$

and where the non-perturbative p_T relative to the parent parton has been neglected. Equation (3.1.101) follows from the $q\bar{q}g$ formula (3.1.11), transformation of variables and the application of the impulse approximation similar to (3.1.95).

For $p_T^2 \to 0$ and $W \to \infty$, p_T fixed, (3.1.101) reduces to

$$\frac{1}{\sigma}\frac{d\sigma}{dxdp_T^2} \simeq \frac{\alpha_s}{2\pi}C_F\frac{|\ln(p_T/xW)|}{p_T^2} [D_q(x)+D_{\bar{q}}(x)] \qquad (3.1.102)$$

which shows that the cross section behaves like $|\ln p_T|/p_T^2$ for p_T small. This is equivalent to the approximation (3.1.19) in case of the thrust distribution. Of course, (3.1.102) is not applicable where the logarithm becomes large. Similar calculations for two-particle cross sections were done by Schierholz and Willrodt /1980/, and Mursula /1980/.

Equation (3.1.101) has some interesting qualitative features. Disregarding the W^2 dependence of the fragmentation functions, it leads to the scaling laws

$$\frac{1}{\sigma}\frac{d^2\sigma}{dxdp_T^2} = \frac{\alpha_s(q^2)}{p_T^2} f(x,x_T)$$

$$\frac{1}{\sigma}\frac{d\sigma}{dp_T^2} = \frac{\alpha_s(q^2)}{p_T^2} g(x_T) \quad . \tag{3.1.103}$$

For $p_T^2 \to 0$ these cross sections have the characteristic behaviour p_T^{-2} originating from the $q\bar{q}g$ cross section (3.1.11). The singularity p_T^{-2}, of course, comes from the infrared singularities in (3.1.11). Therefore (3.1.101) should be applied only for larger p_T's. The formulae (3.1.103) are to be compared to

$$\frac{1}{\sigma}\frac{d^2\sigma}{dxdp_T^2} = \frac{c}{p_T^4} h(x,x_T) \tag{3.1.104}$$

[c has the dimension $(\text{mass})^2$] for the higher-twist contributions which come from diagrams in Fig.3.27 and an exponential decrease in p_T from the non-perturbative jet spread. It would be nice to establish the $1/p_T^2$ fall-off for fixed x, x_T which really will tell, how much of the jet broadening is due to gluon bremsstrahlung and how much to higher-twist contributions. This requires, of course, to compare data at different W^2. On the theoretical side it is difficult to calculate c in (3.1.104). Estimates by Grayon and Tuite /1982/ indicate that c is small, so that higher-twist terms are

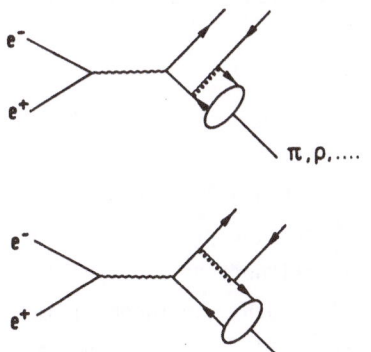

Fig.3.27. Higher-twist contribution to e^+e^- annihilation into hadrons

negligible at PETRA energies (see also Berger /1980/). It was also shown by Söding /1981/ that the p_T^2 distributions cannot be fitted by higher twist terms alone. The W^2 dependence of the data between 12 GeV and 33 GeV differs completely from that given by the higher twist terms. The W dependence of the p_T^2 distribution appears to clearly call for a pointlike contribution.

Of course, the interesting region is large p_T^2. This can be reached only with large x as one can see from Fig.3.28, where (3.1.101) has been plotted for special x values. So far, there are data only for $d\sigma/dp_T^2$. But the average p_T^2 has been measured as a function of x. Hard gluon bremsstrahlung must be reflected in this quantity too. It is defined by

$$<p_T^2(x)> = \int dp_T^2 p_T^2 \frac{1}{\sigma} \frac{d^2\sigma}{dxdp_T^2} \bigg/ \frac{1}{\sigma}\frac{d\sigma}{dx} \quad . \tag{3.1.105}$$

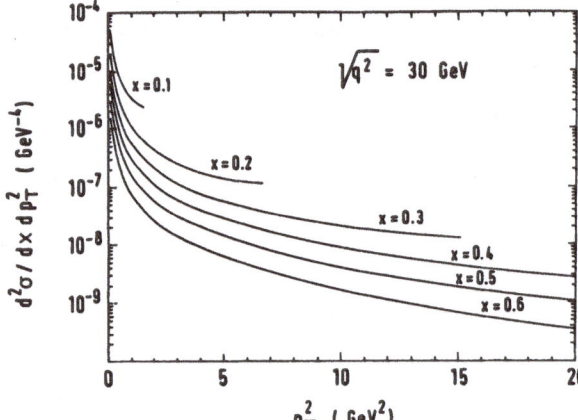

Fig.3.28. Double differential cross section $d^2\sigma/dxdp_T^2$ for various x values

Unlike $d\sigma/dp_T^2$ at very large p_T, this also includes the non-perturbative $q\bar{q}$ background and broadening of the perturbative average p_T^2 caused by fragmentation of quarks and gluons. But for large W^2 and medium x we expect (3.1.105) to be largely determined by single gluon bremsstrahlung which grows like /Politzer, 1977; Kramer and Schierholz, 1979/:

$$<p_T^2(x)> \sim \alpha_s(W^2)W^2 \quad . \tag{3.1.106}$$

Substituting (3.1.101) into (3.1.105) we obtain the curves in Fig.3.29. The fragmentation functions were taken from (3.1.98) and (3.1.99). For two energies 22 and 34 GeV the QCD predictions are compared with recent TASSO data /Althoff et al., 1983/. In these data p_T is measured with respect to the thrust axis. Both, the theoretical curve and the experimental data show the characteristic "seagull" structure, and, as

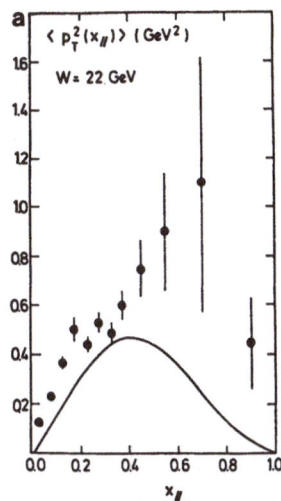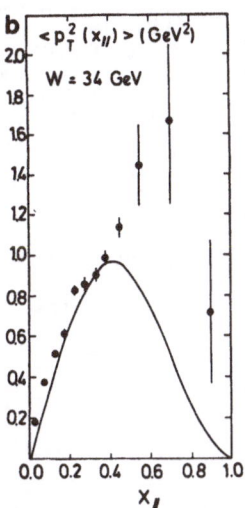

Fig.3.29. The average squared transverse momentum $\langle p_T^2(x) \rangle$ of hadrons measured by TASSO /Althoff et al., 1983/ compared to QCD model based on $q\bar{q}g$ final state; a) 22 GeV and b) 34 GeV

far as one can tell, there is reasonable agreement at both energies for x < 0.4. This agreement is more or less fortuitous as will be explained below. What is more important, is the trend of the data which is much in favour of a rising $\langle p_T^2(x) \rangle$ in agreement with (3.1.106).

The comparison of (3.1.105) together with (3.1.101) with the TASSO data is not realistic. First the data contains the non-perturbative contribution to $\langle p_T^2(x) \rangle$ coming from the 2-jet production. This contribution is approximately equal to the data at lower energies not shown here. Second the formula (3.1.101) is too simple. It neglects the fact that the hadrons are emitted from quark and gluons with a finite non-perturbative p_T. Furthermore (3.1.101) should be used in (3.1.105) only for the larger p_T's, which means that in (3.1.105) the $q\bar{q}g$ contribution comes in only for $p_T^2 \geq (p_T^2)_{min}$ with small p_T^2 contribution replaced by the 2-jet term. All these effects are well accounted for in the Monte-Carlo programs for calculating the fragmentation of quarks and gluons of Sect.2.4.2. In these models the 2- and 3-jet region are separated by a cut-off parameter $y = 1-T_0$. In the 2-jet region the distribution in x and p_T^2 results from the Field-Feynman cascade of q and $\bar{q}$ and in the 3-jet region from the fragmentation of q, $\bar{q}$ and g. This way the $q\bar{q}g$ contribution is reduced appreciably at lower energies and appears only at higher energies where large p_T's are kinematically possible. The parameter y acts like a cut-off in p_T^2 for the 3-jet part. Some examples for seagull structure in the Ali model are exhibited in Fig.3.30 compared to older TASSO data /Brandelik et al., 1979/. In this comparison the $\langle p_T^2(x) \rangle$ of the heavy and of the light jet were plotted sepa-

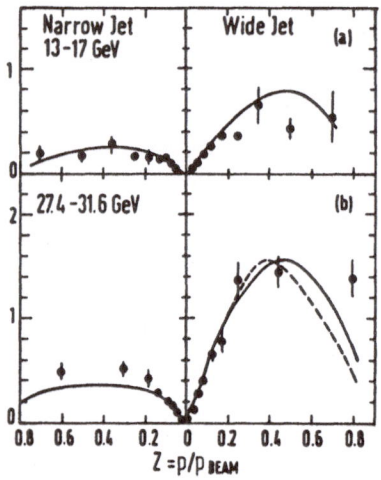
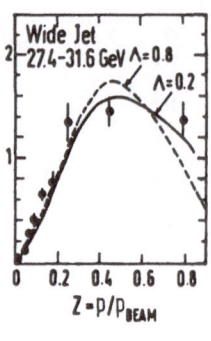

Fig.3.30. Average squared transverse momentum $\langle p_T^2(Z)\rangle$ as a function of the normalized hadron momentum $Z = 2p/W$ ($2p_{beam} = W$) for wide and narrow jet; a) low and b) high energies. In b) the full curve is for gluon fragmentation $D_g(x) \sim x^2+(1-x)^2$ and the dashed curve is for $D_g(x) \sim x(1-x)$. In the right part the curves are for two different couplings α_s: $\Lambda = 0.2$ GeV (full curve) and $\Lambda = 0.8$ (dashed curve) for $D_g \sim x^2+(1-x)^2$. Data are from TASSO. Curves calculated with the Ali model

rately. In the light jet (also called slim jet) we have practically only the non-perturbative $p_T^2(x)$ of the $q\bar{q}$ fragmentation whereas in the heavy jets (also called broad jet) the contribution from the $q\bar{q}g$ intermediate state dominates. This asymmetry of the "seagull" is clearly visible and is well accounted for by the model /Ali et al., 1979,1980; Kramer, 1980/. Of course, part of the asymmetry has its origin in the selection in heavy and light jet as is seen from the curves for energies 13 - 17 GeV. In the calculations two models for the gluon fragmentation into $q\bar{q}$ were used: a) $D_g(x) \sim [x^2+(1-x)^2]$ and b) $D_g \sim x(1-x)$. Both give similar results. In the lower part of Fig.3.30 the variation of the "seagull" with α_s has been studied. We see varying $\alpha_s \sim 1/\ln(q^2/\Lambda^2)$ with Λ between 0.2 and 0.8 GeV has not a dramatic effect on $\langle p_T^2(x)\rangle$. Therefore the seagull structure is not very useful for determining α_s. Similar comparisons have been made also by other groups /Hoyer, Sander, Walsh and Zerwas, 1979; Berger et al., 1979; Brandelik et al., 1979/. With the fragmentation models of Sect.2.4.2 available it is no major problem to compute the distributions $d^2\sigma/dxdp_T^2$ and $d^2\sigma/dp_T^2$ and compare to experimental data. This has been done for the latter distribution. We show in Fig.3.31 the distribution in $p_{T\ out}^2$ and $p_{T\ in}^2$ for lower and higher energies (W = 12 GeV and $27.4 \leq W \leq 36.6$ GeV) /Brandelik et al., 1979; Wolf, 1981/. Here $p_{T\ in}$ is the transversal momentum of charged hadrons in the event plane defined by the sphericity tensor, $p_{T\ out}$ is the transversal momentum with respect to this plane. The increase of the average $p_{T\ in}^2$ with increasing

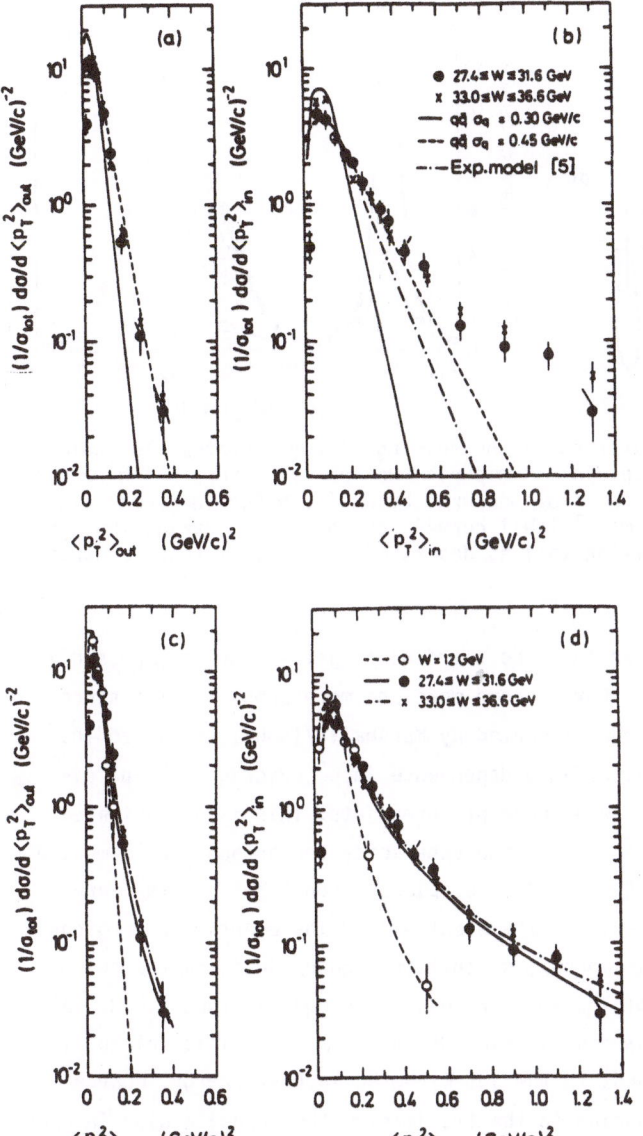

<u>Fig.3.31.</u> Distribution of the mean squared transverse momentum per event with respect to the jet axis, normal to and in the event plane as measured by TASSO for W's between 12 and 36.6 GeV compared to ad-hoc jet broadening models without planar structure [a) and b)] and with QCD (Ali model) in c) and d)

W is evident. Furthermore it is shown, that models with $q\bar{q}$ production only, but with a larger non-perturbative p_T^2 (σ_q = 0.45 GeV and the exponential model), i.e. without the third jet responsible for the planar structure, cannot explain the data. The p_T^2 out distribution is accounted for, but not the p_T^2 in distribution. One also sees quite nicely that the p_T^2 in and p_T^2 out distributions are very similar at 12 GeV where $q\bar{q}g$ effects are hidden underneath the $q\bar{q}$ production. All distributions are well described with the Ali model. More elaborate comparisons of various distributions with this model are found in the work of Brandelik et al. /1980a/.

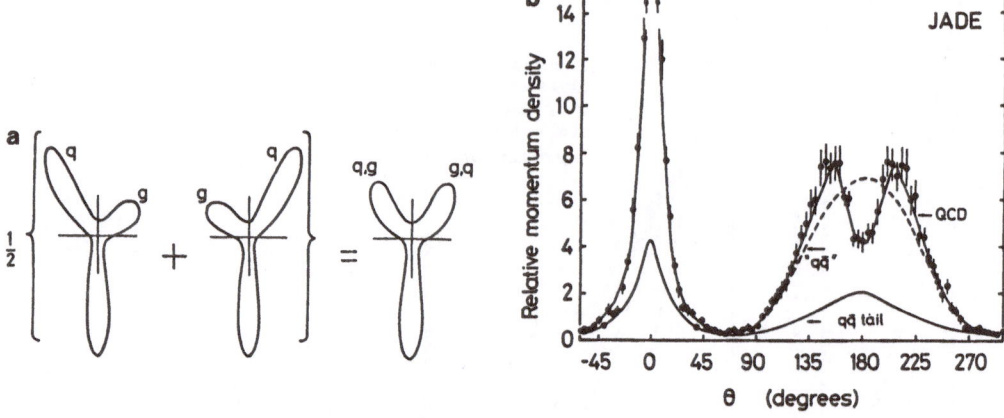

Fig.3.32. a) Diagrammatic representation of the folding of two momentum flow polar plots for the final state qq̄g to obtain a symmetric two lobe structure which recoils against the narrowst jet. b) Momentum flow measured by JADE showing two symmetric lobes at 180° compared to qq̄ + qq̄g model (full curve), qq̄ component leaking through and "qq̄" model with long fragmentation tail as described in the text (dashed curve)

Such demonstrations which test whether the third gluon jet is the only possibility to explain high energy e^+e^- data have become more and more sophisticated since 1979. A further nice example has been presented by Marshall /1980/. He showed that a particular ad hoc model with well-defined dependence in p_T^2 describes the p_T^2 distributions in Fig.3.31 but not the occurrence of three jets. This is demonstrated in Fig.3.32. Experimentally for every event the sphericity tensor has been computed. Then events with $Q_1 < 0.06$ and $Q_2-Q_1 > 0.07$, i.e. planar, non-2-jet events, were selected. $\theta = 0$ is the sphericity axis in the event plane. The events from regions $0 \leq 0 \leq \pi$ and $\pi \leq \theta \leq 2\pi$ were superimposed, so that the "quark jet" and the "gluon jet" were not distinguished and both regions are populated equally (see Fig.3.32a). In Fig.3.32b we see the energy-flow distribution dE/dθ in the sphericity plane. The maximum of dE/dθ at $\theta = 0$ corresponds to the jet with the largest energy. Near $\theta = \pi$ we see the double maximum corresponding to the two jets on the opposite side in good agreement with QCD expectations. The alternative to this is the model with two broad jets only with the following p_T distribution with respect to the jet axis

$$\frac{d\sigma}{dp_T^2} = \frac{1}{\sigma_q} e^{-p_T^2/2\sigma_q} + \frac{0.8}{m} \theta(p_T - \frac{m}{2})\theta(W-m) \tag{3.1.107}$$

and $\sigma_q = 0.33$ GeV and m = 7 GeV. This ansatz reproduces the p_T^2 distribution in Fig. 3.31 because of the second term in (3.1.107).But this model does not explain the 3-jet structure in dE/dθ. It corresponds to the dotted curve ("qq̄") in Fig.3.32b. The usual qq̄ contribution (qq̄ tail) is also plotted. This, of course, reproduces neither the normalization nor the 3-jet structure.

As the last point of this section we shall consider the confrontation of the energy-energy correlations derived in Sect.3.1.5 with experiment. A first experimental study of the energy weighted angular correlation and a comparison with theoretical models have been published by the PLUTO Collaboration at PETRA /Berger et al., 1980b,1981b/. New analyses have been recently reported by the CELLO Collaboration at PETRA /Behrend et al., 1982b/ the MARK II Collaboration /Schlatter et al., 1982/ and the MAC Collaboration /MAC Collaboration, 1982/ at PEP and the MARK J Collaboration at PETRA /MARK J Collaboration, 1982; Adeva et al., 1983c/. Experimentally the energy-energy correlation function is obtained by measuring

$$\frac{1}{\sigma_{tot}} \frac{d\Sigma}{d\Omega d\Omega'} = \frac{1}{N} \frac{1}{\Delta\Omega} \frac{1}{\Delta\Omega'} \sum\sum \frac{EE'}{W^2} \qquad (3.1.108)$$

where σ_{tot} is the total hadronic cross section and E and E' are the energies of the particles in the solid angles $\Delta\Omega$ and $\Delta\Omega'$, respectively. The first sum is over all N events and the second sum is over all pairs of particles in $\Delta\Omega$ and $\Delta\Omega'$. To compare with the corresponding theoretical prediction all angles are summed over except the angle χ between $\Delta\Omega$ and $\Delta\Omega'$ as shown in Fig.3.33. The resulting cross section is

$$\frac{1}{\sigma_{tot}} \frac{d\Sigma}{d\cos\chi} = \frac{1}{N} \frac{1}{\Delta\cos\chi} \sum\sum \frac{EE'}{W^2} \quad . \qquad (3.1.109)$$

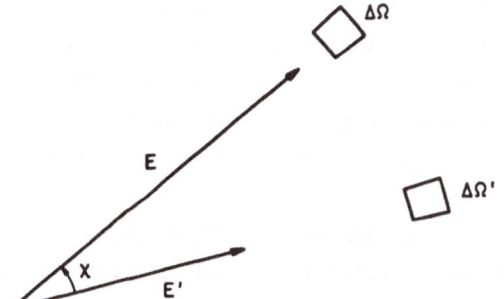

Fig.3.33. Variables for energy-correlation function

This is the analogous definition as (3.1.35) for the energy-energy correlation for parton pairs in QCD. In the definition of neither the theoretical cross section (3.1.35) nor the experimental one (3.1.109) a determination of jet axis or a selection of special classes of events, i.e. two jets, three jets, four jets etc., is required. This may seem to be an advantage. But actually it is not. In the perturbative cross section (3.1.35) the final state $e^+e^- \to q\bar{q}$ does not contribute for $\chi \neq 0,\pi$ so that outside the forward and backward direction the whole cross section results from the $q\bar{q}g$ final state. In the energy-energy correlation for hadrons, however,

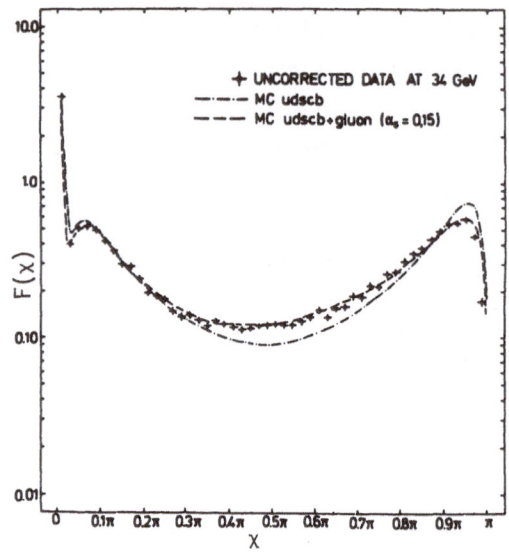

Fig.3.34. Observed energy correlation function $F(\chi)$ as a function of χ at 34 GeV measured by CELLO. Dashed line gives $q\bar{q} + q\bar{q}g$ prediction with Hoyer model ($\alpha_s = 0.15$) and dashed dotted line is for $q\bar{q}$ model with fragmentation

$e^+e^- \to q\bar{q} \to$ hadrons contributes for all angles χ between 0 and π and even at $\chi = \pi/2$, where the dominant process is expected to be the emission of an additional hard gluon, the largest fraction of the cross section originates from $e^+e^- \to q\bar{q}$ with subsequent fragmentation into hadrons. This is seen quite clearly in Fig.3.34 where recent data of the CELLO Collaboration are compared with the Field-Feynman model (MC udscb in the figure) and with the prediction based on the Hoyer model with $\alpha_s = \pi/2$. For $\chi = \pi/2$ the additional $q\bar{q}g$ contribution leads to an increase of the cross section by roughly 30% as compared to the $q\bar{q}$ component. Agreement with the data is reached only with the $q\bar{q} + q\bar{q}g$ model with a coupling constant α_s of the same order as obtained in the other analysis. One should notice that in Fig.3.34 $F(\chi)$, equal to $(1/\sigma_{tot})d\Sigma/d\chi$, is plotted as a function of χ.

These conclusions must also be drawn from recent results of the MARK II and MAC Collaboration at PEP. They fitted the energy-energy correlation of (3.1.109) with the following ansatz

$$\frac{1}{\sigma_{tot}}\frac{d\Sigma}{d\cos\chi} = \alpha_s F_{QCD}(\chi) + \frac{A_0}{W\sin^3\chi} + \frac{\alpha_s A_1}{W}\begin{cases} \sin^3\chi & \chi \leq \frac{\pi}{2} \\ 1+\cos\chi \quad \text{MARK II} \\ \sin\chi \quad \text{MAC} \end{cases} \Bigg\} \quad \chi \geq \frac{\pi}{2} \qquad (3.1.110)$$

In (3.1.110) the first term is the perturbative contribution (3.1.35). The second term is supposed to simulate the non-perturbative $q\bar{q}$ component which is assumed to be symmetrical under the exchange $\chi \to \pi-\chi$ as one would expect. The third terms stands for fragmentation effects on the $q\bar{q}g$ component which is asymmetrical. The results of the fit to the MAC data is shown in Fig.3.35 giving $\alpha_s = 0.20 \pm 0.01 \pm 0.02$, $A_0 =$

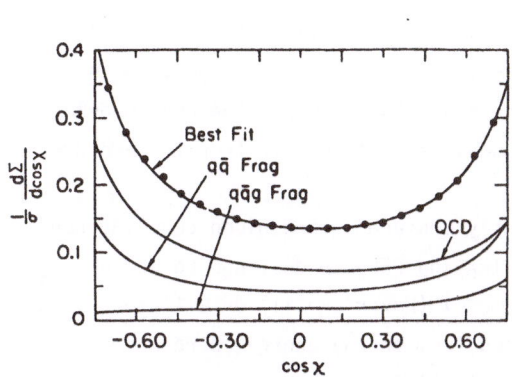

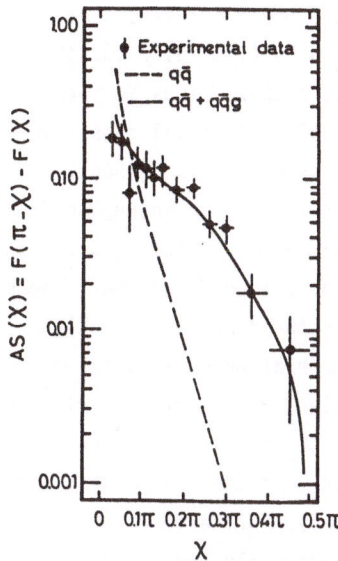

Fig.3.35. Experimental energy correlation as a function of cosχ together with fitted QCD prediction q$\bar{q}$ and q$\bar{q}$g fragmentation terms. Data from MAC Collaboration at PEP ring

Fig.3.36. Asymmetry distribution AS(χ) at 34 GeV corrected to the level of final state particles (no decays) as measured by CELLO compared to QCD prediction with Hoyer model (full curve). Dashed curve is for qq component only

$(1.2 \pm 0.08 \pm 0.15)$ GeV and $A_1 = (2.5 \pm 0.2 \pm 0.4)$ GeV. The corresponding numbers for the MARK II fit are: $\alpha_s = 0.19 \pm 0.02$, $A_0 = (0.7 \pm 0.2)$ GeV and $A_1 = (2.6 \pm 0.5)$ GeV. In Fig.3.35 we see again that the q$\bar{q}$ component makes a substantial contribution. Although the ansatz (3.1.110) seems rather ad-hoc it again leads to reasonable results for α_s. Since the term proportional to A_1 is present it is also not possible to determine α_s from the QCD asymmetry AS(cosχ) [see (3.1.43)] without inducing fragmentation effects. The asymmetry AS(χ) = F(π-χ)-F(χ) derived from the CELLO data in Fig.3.34 is plotted in Fig.3.36 together with the contribution coming from q$\bar{q}$ alone and the model prediction based on the Hoyer model with $\alpha_s = 0.15$. We see that the q$\bar{q}$ component is very much reduced and becomes negligible for $\chi > 0.2\pi$. It should be stressed however that the dynamics of the fragmentation of q$\bar{q}$g can also contribute so that the perturbative AS(χ) given by (3.1.43) does not fully describe the data.

Whereas the formula (3.1.110) used by the PEP groups has certainly the advantage that is does not need long model calculations, the division into the three terms in (3.1.110) has not been tested yet. For this one would need data for various W's in order to separate the fragmentation contributions, which behave like 1/W. On the other hand the CELLO data were analysed originally only with the Hoyer model. The analysis was repeated with the Lund fragmentation scheme. The result is reported in a recent paper by Behrend et al. /1983/. They quote α_s = 0.25 ± 0.04 for the Lund model and α_s = 0.15 ± 0.02 for the Hoyer model from a fit to AS(χ) at W = 34 GeV (see also Ellis /1982/). Thus it seems that at present energies also the energy-weighted asymmetry AS(χ) is still sensitive to the type of correlation induced by the fragmentation of quarks and gluons.

Besides the energy-energy correlations the CELLO Collaboration /Behrend et al., 1983/ analysed also other quantities concerning their sensitivity on the fragmentation model. They found that, depending on the distribution used, the Lund model gives α_s between 28% and 52% higher than the model of Hoyer et al. (for the energy-energy correlation it is 67%).

In Table 3.2 we give a compilation of recently measured quark-gluon coupling constant α_s, obtained by the various groups at PETRA and PEP. We distinguish whether α_s is deduced with the independent fragmentation model (Hoyer or Ali) or with the string model (Lund). Furthermore the α_s values obtained from energy-energy correlation measurements are collected separately. The average of all these values is α_s = 0.19 ± 0.04. This corresponds to a Λ value of $(0.46 \,^{+\,0.50}_{-\,0.31})$ GeV based on the first order formula $\alpha_s = 12\pi/[(33-2N_f)\ln W^2/\Lambda^2]$ with N_f = 5.

In conclusion we can state that there is no alternative to QCD. All tests have shown that the experimental data on hadron production in high energy e^+e^- annihilation can be understood only in terms of the $q\bar{q}g$ model with a vector gluon g or further improvements on it which will be considered in the next section.

Table 3.2. Strong coupling constant α_s obtained by various collaborations by fitting their data with the independent fragmentation or the string fragmentation model to jet distributions and energy-energy correlations in first order QCD. References: Behrend et al. /1983/, Bartel et al. /1982/, MAC /1982/, Schlatter et al. /1982/, MARK J /1982/, Berger et al. /1980/ and Wu /1983/

Collaboration	Indep. Frag.	String	E-E Correlation
CELLO	0.155 - 0.20	0.235 - 0.28	0.15 - 0.25
JADE	0.20 ± 0.015 ± 0.03		
MAC			0.20 ± 0.01 ± 0.02
MARK II			0.19 ± 0.02
MARK J			0.16 ± 0.19
PLUTO	0.15 ± 0.02 ± 0.03		
TASSO	0.194 ± 0.005 ± 0.03		

3.2 Jet up to Order α_s^2

3.2.1 Introduction

In this section we shall discuss all effects which appear if we go one order higher in the QCD perturbation theory, i.e. up to order g^4. Then one more gluon can be emitted from the quark legs. This means we have 4 partons in the final state which will be interpreted as 4-jet production. These 4-parton final states are (i) $e^+e^- \to q\bar{q}gg$ and (ii) $e^+e^- \to q\bar{q}q\bar{q}$, where the second $q\bar{q}$ pair may differ in flavour from the first one. The complete list of diagrams for these two processes is shown in Fig.3.37. Their contribution to the e^+e^- annihilation cross section has been computed by several groups: Ali et al. /1979,1980/; Körner, Schierholz and Willrodt /1981/; Gaemers and Vermaseren /1980/; Nachtmann and Reiter /1982a,1982b/. Their contribution to 4 jets in the final state will be the subject of Sect.3.2.3. Similar to the case of single bremsstrahlung also for double bremsstrahlung one gluon can be emitted almost collinear with a quark or a gluon, or a gluon can be emitted with a very small energy. Also two quarks can come out collinear or soft. All these configurations contribute to the 3-jet cross section and must be considered separately. Similarly with two pairs of quarks and gluons collinear or both gluons soft we have essentially 2-jet configurations. Therefore the diagrams in Fig.3.37 with 4 partons in the final state yield only in kinematic regions away from these degenerate regions, with one or two parton pairs collinear or one or two partons soft, genuine 4 jets. The degenerate regions correspond to 3 and 2 jets. Of course, if we integrate over these degenerate regions we encounter the familiar infrared and mass singularities which produce terms proportional to ε^{-2} and ε^{-1} in the 3-jet contributions and terms proportional to ε^{-k} (k = 1, 2, 3, 4) in the 2-jet contributions ($2\varepsilon = 4-n$ with n being the arbitrary dimension in the dimensional regularization method). The singular terms in the 3-jet cross section are cancelled against the singular terms originating from $O(\alpha_s^2)$ virtual corrections to 3 jets. They result from the Feynman graphs in Fig.3.38 (of

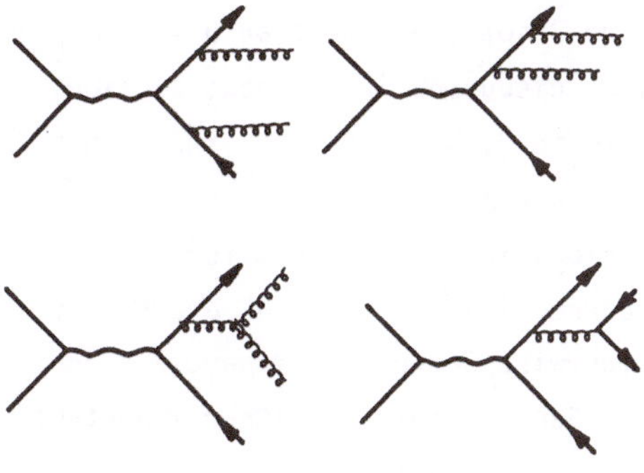

Fig.3.37. Diagrams with four partons ($q\bar{q}qg$ and $q\bar{q}q\bar{q}$) in the final state

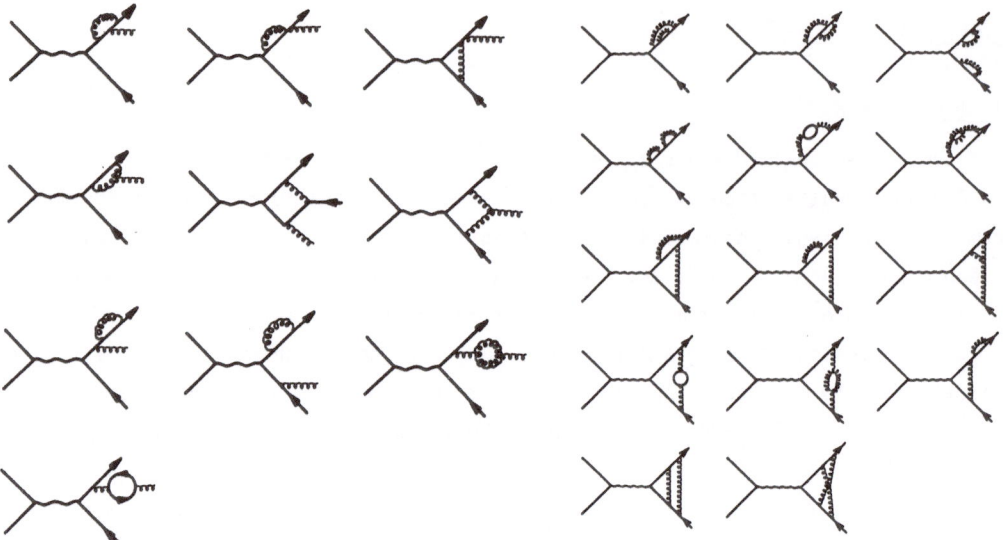

Fig.3.38. Diagrams with $q\bar{q}g$ in the final state to order α_s^2 interfering with the diagrams in Fig.3.1b

Fig.3.39. Diagrams with $q\bar{q}$ in the final state to order α_s^2 interfering with the diagram in Fig.2.1

order g^3) which are multiplied with the $q\bar{q}g$ graphs of order g. The remaining terms in the sum of 4-parton terms integrated over the singular region and the virtual $q\bar{q}g$ contributions yield the $O(\alpha_s^2)$ corrections to the 3-jet cross section. The contributions ε^{-k} (k = 1, 2, 3, 4) from the 4-parton diagrams are supposed to cancel if the following 2-jet contributions are added: (i) the 2-jet contribution in the virtual $q\bar{q}g$ diagrams of Fig.3.38, (ii) the virtual two-loop corrections to the $q\bar{q}$ final state. The second class consists of the $O(g^4)$ graphs in Fig.3.39 which are multiplied with the $q\bar{q}$ graph in lowest (g^0) order and the product of the diagrams in Fig.3.1a being $O(g^2)$. The sum of all these contributions yields the $O(\alpha_s^2)$ correction to the 2-jet cross section. The calculation of these contributions to 2 jets has not been finished yet. It is particularly complicated since all terms from ε^{-4} to ε^0 must be computed. The $O(\alpha_s^2)$ corrections to the 3-jet cross section $\sigma = \sigma_U + \sigma_L$ have been completely calculated. The results will be discussed in the next section. We conclude that the diagrams in Fig.3.37 contribute to 2, 3 and 4 jets, those in Fig.3.38 to 2 and 3 jets and those in Fig.3.39 only to 2 jets.

The $O(\alpha_s^2)$ corrections to 2- and 3-jet cross sections are of interest for several reasons. First we would like to know whether the corrections of order α_s^2 are really small as compared to the order α_s contribution. Otherwise we would not have a "convergent" QCD perturbation theory for jet phenomena. Second, this is rather important, it is well known, that the coupling constant α_s is not uniquely defined. The definition of α_s depends on the renormalization scheme. This renormalization of the quark-gluon coupling constant appears the first time in connection with the virtual diagrams of order α_s^2. This means, only in order α_s^2 it is known which definition of the coupling constant has been used. This is very important if we want to compare the values for α_s deduced from different quantities in e^+e^- annihilation or even from different processes. The same remarks apply concerning the scale at which $\alpha_s(q^2)$ is defined. This may be q^2 or some other scale. Only in order α_s^2 it is possible to find out which scale makes the perturbation theory optimal, i.e. produces the smallest correction terms for a number of physical quantities. Third, QCD shows its full gauge structure only in second order — or higher-order perturbation theory (order $\geq \alpha_s^2$), where the triple-gluon coupling comes in. In all three classes of diagrams, Figs. 3.37-39, we have diagrams with the 3-gluon coupling. In order α_s, however, all diagrams in QCD and in an abelian vector-gluon theory are identical. The only difference is a rescaling of the coupling constant, since $C_F = 1$ in the abelian theory.

3.2.2 Three-Jet Cross Section up to $O(\alpha_s^2)$

To obtain the $O(\alpha_s^2)$ corrections to the 3-jet cross section, first, one calculates the contribution of the virtual diagrams in Fig.3.38. They depend on the same kinematic variables as the first order cross section (3.1.11), namely x_1 and x_2 (or $x_{13} = 1-x_2$ and $y_{23} = 1-x_1$) and contain infrared and mass singularities proportional to ε^{-2} and ε^{-1}. The calculations are rather involved and could be managed only with the help of algebraic computer programs. The final result has the relatively compact form /Ellis, Ross and Terrano, 1981; Fabricius, Kramer, Schmitt and Schierholz, 1982; Lampe and Kramer, 1983/:

$$\frac{d^2\sigma}{dx_1 dx_2} = \sigma^{(2)} \frac{\alpha_s(\mu^2)}{2\pi} C_F \left(\frac{4\pi\mu^2}{q^2}\right)^\varepsilon \frac{1}{\Gamma(1-\varepsilon)} [(1-x_1)(1-x_2)(1-x_3)]^{-\varepsilon} T(x_1,x_2) \quad (3.2.1)$$

where

$$T(x_1,x_2) = \frac{\alpha_s(\mu^2)}{2\pi} \left(\frac{4\pi\mu^2}{q^2}\right)^\varepsilon \frac{\Gamma(1-\varepsilon)}{\Gamma(1-2\varepsilon)} \left\{ B(x_1,x_2)\left[-\frac{1}{\varepsilon^2}(2C_F+N_c)+\frac{a}{\varepsilon}+b\right] + f(x_1,x_2)\right\}. \quad (3.2.2)$$

$B(x_1,x_2)$ was defined in (3.1.62). It depends also on ε. a, b and $f(x_1,x_2)$ have the form:

$$a = -3C_F - (\frac{11}{6}N_c - \frac{1}{3}N_f) + (2C_F - N_c)\ln y_{12} + N_c\ln(y_{13}y_{23}) \quad (3.2.3)$$

$$b = (2C_F + N_c)\frac{\pi^2}{3} - 8C_F + \frac{1}{2}N_c(\ln^2 y_{12} - \ln^2 y_{13} - \ln^2 y_{23}) - C_F \ln^2 y_{12}$$

$$- (\frac{11}{6}N_c - \frac{1}{3}N_f)\left(\gamma - \ln\frac{4\pi\mu^2}{q^2}\right) \tag{3.2.4}$$

$$f(x_1, x_2) = \left(C_F\frac{y_{12}}{y_{12}+y_{23}} + \frac{y_{12}}{y_{12}+y_{13}} + \frac{4y_{12}}{y_{13}+y_{23}} - \frac{y_{12}}{y_{13}} - \frac{y_{12}}{y_{23}} - \frac{y_{13}}{y_{23}} - \frac{y_{23}}{y_{13}}\right)$$

$$+ N_c\left(\frac{y_{12}}{y_{13}} + \frac{y_{12}}{y_{23}} + \frac{y_{13}}{y_{23}} + \frac{y_{23}}{y_{13}} - \frac{2y_{12}}{y_{13}+y_{23}}\right)$$

$$+ \ln y_{13}\left\{N_c\frac{y_{13}}{y_{12}+y_{23}} + C_F\left(4 - \frac{y_{13}y_{23}}{(y_{12}+y_{23})^2} + \frac{2y_{12}-4y_{23}}{y_{12}+y_{23}}\right)\right\}$$

$$+ \ln y_{23}\left\{N_c\frac{y_{23}}{y_{12}+y_{13}} + C_F\left(4 - \frac{y_{13}y_{23}}{(y_{12}+y_{13})^2} + \frac{2y_{13}-4y_{23}}{y_{12}+y_{13}}\right)\right\}$$

$$+ 2(2C_F - N_c)\ln y_{12}\left[\frac{2y_{12}}{y_{13}+y_{23}} + \frac{y_{12}^2}{(y_{13}+y_{23})^2}\right] - N_c B^v(y_{13}, y_{23})r(y_{13}, y_{23})$$

$$- 2(2C_F - N_c)\left[\frac{y_{12}^2 + (y_{12}+y_{13})^2}{y_{13}y_{23}}r(y_{12}, y_{23}) + \frac{y_{12}^2 + (y_{12}+y_{23})^2}{y_{13}y_{23}}r(y_{12}, y_{23})\right] . \tag{3.2.5}$$

In (3.2.4) γ is the Euler constant. $r(x,y)$ in (3.2.5) is the function

$$r(x,y) = \ln x \ln y - \ln x \ln(1-x) - \ln y \ln(1-y) + \frac{\pi^2}{6} - \mathscr{L}_2(x) - \mathscr{L}_2(y) \tag{3.2.6}$$

with $\mathscr{L}_2(x)$ being the Spence function

$$\mathscr{L}_2(x) = -\int_0^x dz\frac{\ln(1-z)}{z} . \tag{3.2.7}$$

In (3.2.4) we notice several expressions which are proportional to $11N_c/6 - N_f/3$ and, hence, can be absorbed into the definition of the strong coupling constant. The large logarithmic term

$$(\frac{11}{6}N_c - \frac{1}{3}N_f) \ln\frac{\mu^2}{q^2} \cdot B(x_1, x_2)$$

in (3.2.4) represents the explicit beginning of the renormalization group improvement of (3.1.11) [or (3.1.61)] and arranges that $\alpha_s(\mu^2)$ becomes the running coupling constant

$$\alpha_s(q^2) = \alpha_s(\mu^2) \left[1 + \frac{\alpha_s(\mu^2)}{2\pi} \left(\frac{11}{6} N_c - \frac{1}{3} N_f \right) \ln\frac{q^2}{\mu^2} \right]^{-1} \quad . \tag{3.2.8}$$

For better convergence of the perturbation series it is customary to also subtract the expression $(N_f/3 - 11N_c/6)(\gamma - \ln 4\pi)$ from (3.2.4) together with the ultraviolet pole term in the renormalization procedure which then is called the $\overline{MS}$ scheme. The formulae (3.2.1-5) are valid for the so-called minimal renormalization scheme (MS) where only the pole term $\sim \varepsilon^{-1}$ is subtracted in the renormalization of α_s. We hope that these elucidations sufficiently explain what we meant with renormalization and scale dependence of α_s and how these are fixed in order α_s^2 through the virtual corrections. Some further details will be discussed later in Sect.3.2.4.

The contributions to 3 jets contained in the 4-parton diagrams of Fig.3.37 can be calculated only if parameters are introduced which define the 3-jet region inside the 4-parton phase space. This is not unique. The only requirement is that this region contains the pure 3-jet limit. We employ the analogous definitions for separating 3 and 4 jets, which we employed for separating 2 and 3 jets in Sect.3.1.7, the ε, δ boundary of Sterman and Weinberg and the invariant mass boundary. Then with the ε, δ definition we understand by 3-jet cross section now the cross section for events which have all but a fraction $\varepsilon/2$ of the total energy W distributed within three separated cones of (full) opening angle δ. In other words, we call an event (on the parton level) a 3-jet event, if all parton momenta fall inside the phase volume shown in Fig.3.40. This includes the singular region associated with one of the gluons being soft and/or collinear with one of the quarks or the other gluon (in $e^+e^- \rightarrow q\bar{q}gg$) and one of the quarks being collinear with one of the antiquarks (in $e^+e^- \rightarrow q\bar{q}q\bar{q}$), respectively.

The 3-jet cross section is again finite by virtue of the Kinoshita-Lee-Nauenberg theorem, that is to say, the processes $e^+e^- \rightarrow q\bar{q}gg$ and $q\bar{q}q\bar{q}$ must contribute the same pole terms in 4-n = 2ε as the loop corrections (3.2.1-5). After this the 3-jet cross section, which now is

$$d\sigma(\varepsilon,\delta) = d\sigma^{(3)} + d\sigma^{(4)}(\varepsilon,\delta) \quad , \tag{3.2.9}$$

Fig.3.40. Three-jet space volume

where $d\sigma^{(3)}$ includes the $O(\alpha_s)$ contribution (3.1.11) and the $O(\alpha_s^2)$ loop corrections (3.2.1), has in the $\overline{MS}$ renormalization scheme the following form:

$$\frac{1}{\sigma^{(2)}} \frac{d^2\sigma}{dx_1 dx_2} = \frac{\alpha_s(q^2)}{2\pi} C_F \left\{ B^V(x_1,x_2) \left[1 + \frac{\alpha_s(q^2)}{2\pi} (J_1+J_2+J_3) \right] \right.$$

$$\left. + \frac{\alpha_s(q^2)}{2\pi} f(x_1,x_2) \right\} + O(\varepsilon,\delta) \tag{3.2.10}$$

where

$$J_1 = C_F \left[\left(-2\ln\frac{\varepsilon}{x_1} - 2\ln\frac{\varepsilon}{x_2} - 3 \right) \ln\left(\frac{1-\cos\delta}{2}\right) + 4\ln\varepsilon\ln\left(\frac{x_1+x_2-1}{x_1x_2}\right) \right.$$

$$+ 2\left(\frac{\varepsilon}{x_1}+\frac{\varepsilon}{x_2}\right)\ln\left(\frac{1-\cos\delta}{2}\right) + \ln^2\left(\frac{x_1+x_2-1}{x_1x_2}\right) + 2\ln^2 x_1 + 2\ln^2 x_2$$

$$\left. - 3\ln x_1 - 3\ln x_2 - \ln^2(1-x_3) - 2\mathscr{L}_2\left(\frac{x_1+x_2-1}{x_1x_2}\right) - \frac{\pi^2}{3} + 5 \right] \quad , \tag{3.2.11}$$

$$J_2 = N_c \left\{ \left(-2\ln\frac{\varepsilon}{x_3} - \frac{11}{6} \right) \ln\left(\frac{1-\cos\delta}{2}\right) + 2\ln\varepsilon \cdot \left[\ln\left(\frac{x_1+x_3-1}{x_1x_3}\right) + \ln\left(\frac{x_2+x_3-1}{x_2x_3}\right) \right. \right.$$

$$\left. - \ln\left(\frac{x_1+x_2-1}{x_1x_2}\right) \right] + \frac{2\varepsilon}{x_3}\ln\left(\frac{1-\cos\delta}{2}\right) + \frac{1}{2}\ln^2\left(\frac{x_1+x_3-1}{x_1x_3}\right)$$

$$+ \frac{1}{2}\ln^2\left(\frac{x_2+x_3-1}{x_2x_3}\right) - \frac{1}{2}\ln^2\left(\frac{x_1+x_3-1}{x_1x_3}\right) + 2\ln^2 x_3 - \frac{11}{3}\ln x_3 + \frac{1}{2}\ln^2(1-x_3)$$

$$- \frac{1}{2}\ln^2(1-x_1) - \frac{1}{2}\ln^2(1-x_2) + \mathscr{L}_2\left(\frac{x_1+x_2-1}{x_1x_2}\right) - \mathscr{L}_2\left(\frac{x_1+x_3-1}{x_1x_3}\right) - \mathscr{L}_2\left(\frac{x_2+x_3-1}{x_2x_3}\right)$$

$$- \frac{\pi^2}{6} + \frac{137}{18} - \frac{x_3^2}{x_1^2+x_2^2} + \frac{1}{3}\frac{(1-x_1)(1-x_2)}{x_1^2+x_3^2} \right\} \quad , \tag{3.2.12}$$

$$J_3 = \frac{N_f}{2} \left[\frac{2}{3}\ln\left(\frac{1-\cos\delta}{2}\right) + \frac{4}{3}\ln x_3 - \frac{26}{9} + \frac{1}{3}\frac{x_3^2}{x_1^2+x_2^2} \right] \quad . \tag{3.2.13}$$

$f(x_1,x_2)$ is given in (3.2.5) and B^V in (3.1.63): Of course, for SU(3) colour we must introduce $N_c = 3$, $C_F = 4/3$ and the number of flavours N_f depending on the thresholds reached for chosen W.

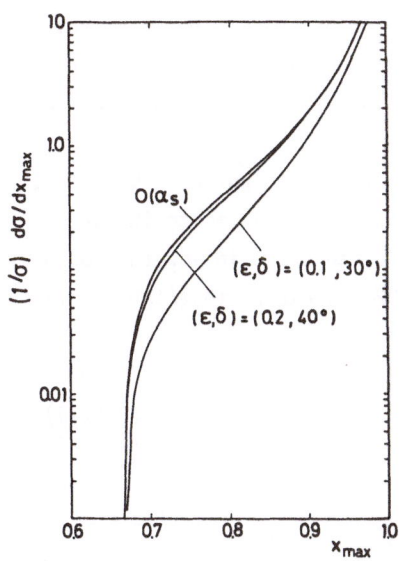

Fig.3.41. Three-jet cross section for $(\varepsilon, \delta) = (0.2, 40°)$ and $(\varepsilon, \delta) = (0.1, 30°)$ with the Born cross section $[O(\alpha_s)]$ as a function of x_{max} for $\alpha_s = 0.16$ and $N_f = 5$

With the formula (3.2.10) /Fabricius, Kramer, Schmitt and Schierholz, 1980,1982; Gutbrod, Kramer and Schierholz, 1983/ the various distributions in one (or two) jet variables can be calculated, similarly, as it was discussed for the low-order cross section in Sect.3.1.3-5. An example is presented in Fig.3.41, where $(1/\sigma)d\sigma/dx_{max}$ is plotted together with $(1/\sigma)d\sigma/dx_{max}$ in Born approximation $[O(\alpha_s)$ curve] for two choices of the pair of parameters ε and δ: $\varepsilon, \delta = 0.2, 40°$ and $\varepsilon, \delta = 0.1, 30°$. The coupling constant $\alpha_s = 0.16$ and $N_f = 5$ are valid for all three curves. x_{max} is the energy of the most energetic jet which equals thrust for three massless partons. σ is equal to the total cross section up to $O(\alpha_s^2)$. The cross section now depends on ε and δ, the resolution parameters for defining 3 jets out of 4 partons. We see that $d\sigma/dx_{max}$ decreases with decreasing ε and δ as we expect from (3.2.10-13). If we choose ε and δ unreasonably small, $d\sigma/dx_{max}$ may become negative. In this region perturbation theory of finite order is not applicable any more and only the summed perturbation series up to finite order would give sensible results. The choice $\varepsilon, \delta = 0.1, 30°$ is already too small, producing $O(\alpha_s^2)$ corrections which changes $d\sigma/dx_{max}$ by a factor of two. In contrast, the choice $\varepsilon, \delta = 0.2, 40°$ is just right. With this the $O(\alpha_s^2)$ corrections are reasonably small compared to the $O(\alpha_s)$ cross section. Thus, with such values of ε and δ we have a reasonable perturbation theory for the 3-jet cross section. A good criterion for the choice of ε and δ is the value of the 2-jet cross section obtained with these parameters. Since $\sigma^{2\text{-jet}}(\varepsilon, \delta)$ has not been computed yet up to $O(\alpha_s^2)$ we use the $O(\alpha_s)$ Sterman-Weinberg formula (3.1.73). This yields for the 2-jet multiplicity 67% if $\varepsilon, \delta = 0.2, 40°$ and 36% if $\varepsilon, \delta = 0.1, 30°$ and $\alpha_s = 0.16$. So the choice $\varepsilon, \delta = 0.1, 30°$ corresponds also to a rather large change of the 2-jet multiplicity compared to the zeroth order value. The exact value of the boundary parameters ε and δ must be chosen in accordance with the experimental data analysis, for example, inside the cluster algorithm as was described in Sect.3.1.9.

The 3-jet cross section has been calculated also for the invariant mass boundary. In this case all configurations of the 4-parton phase space contribute to the 3-jet cross section, for which the four momenta p_i, p_j of two partons are inside the boundary $(p_i + p_j)^2 \leq yW^2$. The result has the same form as (3.2.10) except that the quantities J_1, J_2 and J_3 have now the following form:

$$
J_1 = C_F \left(-2\ln^2 \frac{y}{y_{12}} - 3\ln y - 1 + \frac{\pi^2}{3} + \frac{2y}{y_{12}} \ln \frac{y^2}{y_{12}} \right)
\tag{3.2.14}
$$

$$
J_2 = N_c \left(\ln^2 \frac{y}{y_{12}} - \ln^2 \frac{y}{y_{13}} - \ln^2 \frac{y}{y_{23}} - \frac{11}{6} \ln y + \frac{67}{18} + \frac{\pi^2}{6} \right.
$$
$$
\left. - \frac{y}{y_{12}} \ln \frac{y^2}{y_{12}} + \frac{y}{y_{13}} \ln \frac{y^2}{y_{13}} + \frac{y}{y_{23}} \ln \frac{y^2}{y_{23}} \right)
\tag{3.2.15}
$$

$$
J_3 = \frac{N_f}{2} \left(\frac{2}{3} \ln y - \frac{10}{9} \right) \quad .
\tag{3.2.16}
$$

It should be stressed that the formulae for J_1, J_2 and J_3 in case of the y boundary cannot be derived from (3.2.11-13) by a simple substitution. We notice that the leading logarithmic terms in $\ln\delta^2$, $\ln\varepsilon$ and $\ln y$ agree when we replace $\varepsilon^2 = \delta^2/4$ by y. But the non-leading terms are completely different.

The result (3.2.14-16) may lead us to introduce a new scale for α_s. For example if we change $\alpha_s(q^2)$ in (3.2.10) into $\alpha_s(yq^2)$ we absorb the following term in J_2 and J_3

$$
- \left(\frac{11}{6} N_c - \frac{1}{3} N_f \right) \ln y \, B^V(y_{13}, y_{23}) \quad .
$$

This way the $O(\alpha_s^2)$ correction is diminished by a large term for small enough y. Of course, the resulting α_s determined from some experimental data now will be larger, but it is the α_s for a smaller scale. The choice of the scale is more or less arbitrary. For the "optimal" scale one demands that α_s and the higher order coefficients are small. One possibility could be to choose the scale of α_s in such a way that the higher order contributions vanish. Then all higher order contributions are absorbed in α_s. This defines the so-called "low-order" scheme. Usually this low-order scheme, defined in connection with a specific physical quantity, here the 3-jet cross section, will lead for other physical observables to a "bad" perturbation theory in this coupling, so that there is no advantage employing this particular scheme. We shall make use of this scheme later on in connection with a fit of experimental data to an abelian vector gluon theory. The α_s defined for different scales can be converted into the corresponding Λ values using

$$\alpha_s(q^2) = \frac{2\pi}{(\frac{11}{6}N_c - \frac{1}{3}N_f)\ln\frac{q^2}{\Lambda^2}} \qquad (3.2.17)$$

Then physically equivalent scales should produce roughly the same Λ values.

Finally we remark that J_1 in (3.1.14) agrees in the limit $y_{12} \to 1$ with the factor of $(\alpha_s/2\pi)C_F$ in $\sigma^{2\text{-jet}}_1(y)$ in (3.1.68). This is a necessary condition. Similar tests can be done for J_2 and J_3 in (3.2.15) and (3.2.16) and the J_i for the Sterman-Weinberg cross section.

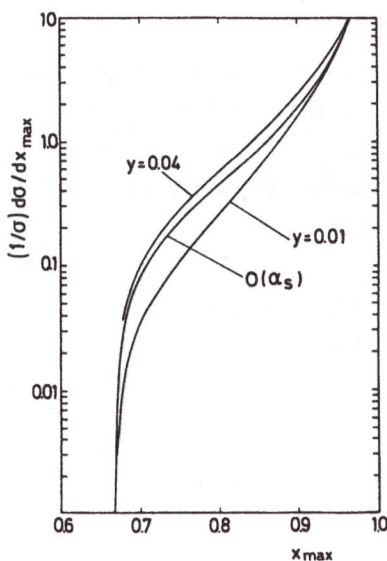

Fig.3.42. Three-jet cross section for $y = 0.04$ and 0.01 with the Born cross section $[O(\alpha_s)]$ as a function of x_{max} for $\alpha_s = 0.16$ and $N_f = 5$

The dependence of $(1/\sigma)d\sigma/dx_{max}$ as a function of y is shown in Fig.3.42 for $y = 0.04$ and 0.01 together with the Born cross section for $\alpha_s = 0.16$ and $N_f = 5$, respectively. For $y = 0.04$ the $O(\alpha_s^2)$ corrected x_{max} distribution differs little from the distribution in Born approximation. Thus also for appropriately chosen y values the α_s^2 corrections to the thrust distribution are reasonably small. In Sect.3.1.7, in connection with the 2-jet cross section in $O(\alpha_s)$, we had found out already that y should lie in the interval $0.03 \leq y \leq 0.05$. The curve for $y = 0.01$, however, deviates appreciably from the lowest order curve. This means that $y = 0.01$ is already outside the range of y values, in which we have a convergent perturbation theory, consistent with the statements made in Sect.3.1.7.

We remark that the curve for y = 0.04 lies somewhat above the $O(\alpha_s)$ curve for all x_{max} values in contrast to the ε, δ = 0.2, 40° curve in Fig.3.41, although $\frac{1}{2}(1 - \cos\delta) \simeq \varepsilon^2 = 0.04$. The reason lies in the different non-leading terms in the two formulae. It means that with the two definitions for jet cross sections, ε, δ or y boundary, completely different regions of the 4-parton cross section are included in 3 jets.

The dependence of $(1/\sigma)d\sigma/dx_{max}$ on y as shown in Fig.3.42 is characteristic for QCD, i.e. for a theory with a non-abelian gluon. It is completely different for an abelian vector gluon theory. In $O(\alpha_s)$ both theories lead to identical predictions except that C_F = 1, which means that the coupling α_A in the abelian theory is related by $\alpha_A = (4/3)\alpha_s$ to the QCD coupling. In $O(\alpha_s^2)$ we have a different behaviour, since the results now depend on two constants C_F and N_c. The predictions of an abelian theory follow from (3.2.10) if we substitute C_F = 1, N_c = 0 and $N_f \rightarrow 2N_f$. The substitution $N_f \rightarrow 2N_f$ comes from the fact that in QCD $Tr(\frac{\lambda a}{2}\frac{\lambda a}{2}) = \frac{1}{2}$ which must be replaced by 1 in the abelian theory. The results are shown in Fig.3.43 for three y values, y = 0.08, 0.04 and 0.02, α_A = 0.21 and N_f = 5. We see that $(1/\sigma)d\sigma/dx_{max}$ decreases much faster with decreasing y than for QCD. The renormalization scheme is the same $\overline{MS}$ scheme as in QCD. This seems an interesting method to verify the non-abelian nature of the gluon and will discussed further when we consider comparisons of the theory with experimental data.

The best way to confront the predictions inherent in the formula (3.2.10) with experimental data is to compare with results of the cluster analysis. In this meth-

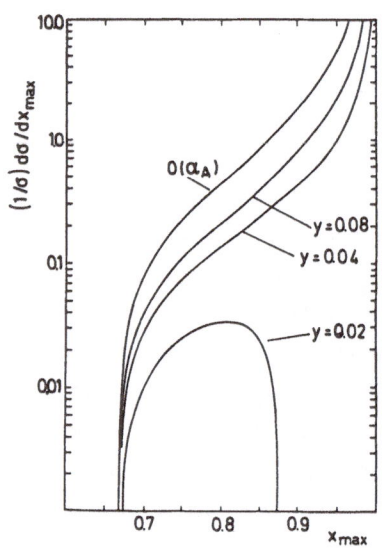

Fig.3.43. Three-jet cross section of abelian vector theory (N_c = 0) for y = 0.08, 0.04 and 0.02 together with Born cross section $O(\alpha_A)$ as a function of x_{max} for α_A = 0.21 and N_f = 5

od, which we explained in some detail in Sect.3.1.9, the hadronization effects are eliminated (at least to a large extent) and the cross section is the 3-jet cross section defined in the same way as in perturbation theory. The parameters ε, δ or y, which were necessary to define the 3-jet cross section in $O(\alpha_s^2)$ perturbation theory, were built into the cluster analysis in an analogous way for separating 3- and 4-cluster events. With this viewpoint the JADE Collaboration at PETRA has performed the cluster analysis, both with the Sterman-Weinberg definition of jets (ε,δ method) and the method based on the invariant mass constraint y /Bartel et al., 1982a/. Their results, both for $(1/\sigma)d\sigma/dx_\perp$ and $(1/\sigma)d\sigma/dx_{max}$ were used to determine the coupling constant α_s by fitting (3.2.10) to their data. One of their examples is shown in Fig. 3.44, where the data for $(1/\sigma)d\sigma/dx_\perp$ and for $(1/\sigma)d\sigma/dx_{max}$ with y = 0.04 can be seen (the y_{max} in the figure is identical to our y). The curves are the $O(\alpha_s)$ prediction (first order contribution) and the prediction for first and second order with the coupling constant α_s = 0.16. For determining α_s only the range $x_{max} \le 0.85$ (x_{max} = x_1 in the figure) and $x_\perp \ge 0.30$ was used, in order to eliminate data points which could be influenced by 2-jet contributions. The fitted values are: α_s = 0.16 ± 0.01 ($x_\perp$ distribution) and α_s = 0.16 ± 0.015 (x_{max} distribution). These values are somewhat smaller than the values fitted to the Born cross section (α_s = 0.20, see Table 3.1). Since the shapes of the curves in first order and in first and second order are very similar both give a good fit to the data. The χ^2/DF values are almost equal for both cases (the first order curve in Fig.3.44 does not fit, because α_s =

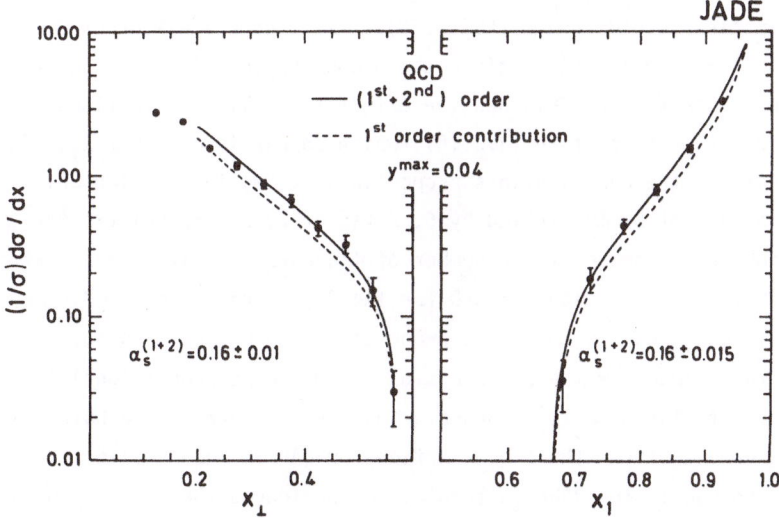

Fig.3.44. Data of the JADE Collaboration obtained with the cluster analysis with y = 0.04 compared to first and second order QCD prediction giving α_s = 0.16 ± 0.01 ($x_\perp$ distribution) and α_s = 0.16 ± 0.015 [x_1 = (x_{max}) distribution]. The dashed curve is the Born cross section with α_s = 0.16

0.16). The α_s values derived with the (ε,δ) method are almost equal: α_s = 0.175 ± 0.012 ($x_\perp$ distribution) and α_s = 0.170 ± 0.010 (x_{max} distribution) written down in Table 3.1. It is very reassuring that the α_s values obtained with the two algorithms for defining jets are so close. The mean value from all these fits is: α_s = 0.165 ± 0.015 (stat.) ± 0.03 (syst.). The systematic error of 0.03 has its origin to a large extent in in-accuracies of the cluster analysis, in particular that one relies on some input from fragmentation models which are not well enough known. This was discussed already in Sect.3.1.9.

The data of the JADE Collaboration have been compared also with an abelian vector gluon theory. Such a theory is, up to the strength of the coupling, equal to quantum electrodynamics, which is a U(1) gauge theory in contrast to the SU(3) gauge theory QCD. The U(1) gauge theory does not have the gluon self coupling, which is character-istic for QCD. As we emphazised already, in order α_s both theories give identical pre-dictions. The formulae for the $q\bar{q}g$ final states and for σ_{tot} are the same, if the quark-gluon coupling of QCD: $C_F\alpha_s$ is replaced by α_A, the coupling of the abelian the-ory, which we shall call QAD in the following. Actually such a theory disagrees with experiment already in zeroth order if we keep the number of flavours the same as in QCD. Since the quarks would have no colour the R value would be a factor of 3 lower. To compensate for this one increases the number of flavours by a factor of three. On-ly in this version of QAD all experimental facts in e^+e^- annihilation can be made to agree with theory in the order α_A by adjusting α_A. This is completely different, when we include also the $O(\alpha_A^2)$ corrections to the 3-jet cross section and to σ_{tot}. In order α_A^2 the contribution of the 3-gluon coupling in the virtual diagrams (Figs. 3.38-39) and in the bremsstrahlung diagrams (Fig.3.37) is missing and all other dia-grams contribute with different weight now, since the gluon couples with the factor 1 instead of the QCD colour matrix $\lambda_a/2$. This has the effect that all contributions pro-portional to N_c in the 3-jet cross section (3.2.10), which appear in J_2 and $f(x_1,x_2)$, must be omitted. In addition, the contributions proportional to N_f in J_3 which ori-ginate from the quark-loops must be multiplied by 6, a factor 2, characteristic for QAD, and the factor 3 for the increase of the number of flavours. We have demonstrated already with Fig.3.43 that the predictions of QAD for the 3-jet cross section differ appreciably from those in QCD, because the terms proportional to N_c are missing. This effect is increased further since now the contribution of the terms proportional to N_f is increased again by a factor of 3 to compensate the quark colour. Since both contributions J_1 and J_3 are negative, the cross section for QAD is much smaller than that for QCD with y chosen equal. The JADE Collaboration considered the case ε = 0.2, δ=60° instead of y=0.04. In order to avoid overestimating the effect of the quark loops the number of flavours has been chosen N_f=3, since the contributions of heavy quarks to the inner loops is damped. They tried to fit the data points in Fig.3.44 by varying α_A.

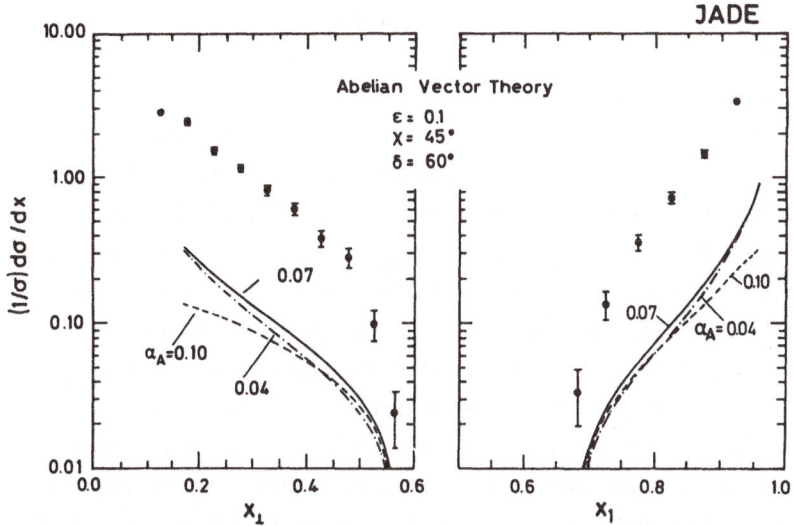

Fig.3.45. Data of the JADE Collaboration obtained for $(\varepsilon,\delta) = (0.2, 60°)$ cuts compared to abelian vector gluon theory in $\overline{MS}$ renormalization scheme for three different coupling constants α_A

However, they did not succeed. Independent of the α_A chosen, the theoretical curve was always at least a factor of 10 below the data points as can be seen in Fig.3.45. The reason for this behaviour is that the $O(\alpha_A^2)$ corrections are already for $\varepsilon = 0.2$, $\delta = 60°$ so strong negative that all theoretical curves lie always below a maximal curve obtained for $\alpha_A = 0.07$ (see Fig.3.45). In this comparison the coupling α_A was defined in the $\overline{MS}$ renormalization scheme. In this scheme the total cross section expressed by R has up to $O(\alpha_A^2)$ the following form

$$R = R_0 \left[1 + \frac{3}{4}\frac{\alpha_A}{\pi} - \left(\frac{3}{32} + 0.519\, N_f\right)\left(\frac{\alpha_A}{\pi}\right)^2 + \ldots \right] \qquad (3.2.18)$$

where R_0 measures the zero-order contribution. The experimental value of R is 3.93 ± 0.10 (see Sect.2.1). From this we determine $(\alpha_A)_{\overline{MS}} = 0.43 \pm 0.22$. Above we have seen that for no value of α_A, i.e. also not for this value, we can get a fit of the data of the 3-jet cross section. Therefore, we conclude that R and the 3-jet distributions $(1/\sigma)d\sigma/dx_{max}$ and $(1/\sigma)d\sigma/dx_\perp$ cannot be explained within the $\overline{MS}$ renormalization scheme of the abelian vector gluon theory, in contrast to QCD, where this is possible without difficulty. We notice that the $(\alpha_A)_{\overline{MS}}$ deduced from R is rather large indicating that a "convergent" perturbation theory may not exist for the $\overline{MS}$ scheme.

Now, the renormalization scheme is arbitrary, as will be explained in more detail in Sect.3.2.4. Then we may ask whether a scheme for the abelian theory exists in which the 3-jet cross sections $(1/\sigma)d\sigma/dx_{max}$ and $(1/\sigma)d\sigma/dx_\perp$ can be fitted. One such scheme is certainly the low-order scheme. We learned in Sect.3.1.9 that both distributions are well described in lowest order with a QCD coupling $\alpha_s = 0.2$ (see Table 3.1). This means that also in QAD a low order fit of both distributions is possible. This scheme has also the advantage that the 4-jet rate, which will be the subject of the next section, can be very well described in the abelian theory. So far we have not made use of any higher order information, which we have on R and the two distributions. For this purpose we fit only one data point for $(1/\sigma)d\sigma/dx_{max}$. We take the point at $x_{max} = 0.825$ which is $(1/\sigma)d\sigma/dx_{max} = 0.72 \pm 0.07$. From this we determine $\tilde{\alpha}_A = 0.24 \pm 0.02$ which is now the value of the coupling in the low-order scheme. With this adjustment of one data point for $(1/\sigma)d\sigma/dx_{max}$ to the low-order formula, clearly it does not follow automatically that the complete $(1/\sigma)d\sigma/dx_{max}$ is given by the low-order formula. But the whole distribution in the low-order scheme can be deduced from (3.2.10). It is shown in Fig.3.46 compared to the data points of Fig.3.45. Compared to the QCD fit the agreement is somewhat worse, none of the other points for $x_{max} < 0.8$ is fitted. But the quality of the fit would not justify to discard the abelian theory. The theory for $(1/\sigma)d\sigma/dx_{max}$ deduced from (3.2.10) gives us also a relation between the coupling $\tilde{\alpha}_A$ in the low-order scheme (defined by a fit to $(1/\sigma)d\sigma/dx_{max}$ at $x_{max} = 0.825$) and the $\overline{MS}$ coupling α_A

$$\frac{\alpha_A}{\pi} = \frac{\tilde{\alpha}_A}{\pi} - 23.2 \left(\frac{\tilde{\alpha}_A}{\pi}\right)^2 \quad . \tag{3.2.19}$$

With this we are able to calculate R up to $O(\tilde{\alpha}_A^2)$ in the low-order scheme. The result for $N_f = 3$ is:

$$R = R_0 \left[1 + \frac{3}{4}\frac{\tilde{\alpha}_A}{\pi} + 15.8 \left(\frac{\tilde{\alpha}_A}{\pi}\right)^2 + \ldots \right] \quad . \tag{3.2.20}$$

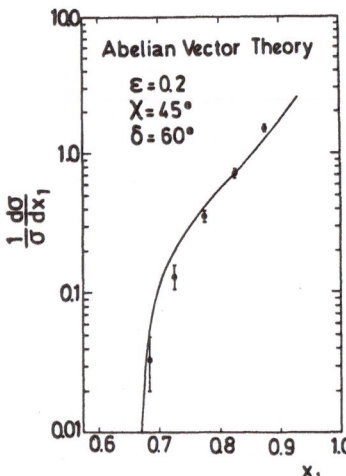

Fig.3.46. JADE data for $(\varepsilon, \delta) = (0.2, 60°)$ compared to abelian vector gluon theory in the low order scheme defined with $(1/\sigma)d\sigma/dx_1$ for $x_1 = 0.825$

For $\tilde{\alpha}_A/\pi = 0.076 \pm 0.008$ we obtain $R = 4.21 \pm 0.09$ in disagreement with the measured value 3.93 ± 0.10. We notice that the $\tilde{\alpha}_A$ scheme produces a relatively bad perturbation series for R. The coefficient of $\tilde{\alpha}_A^2$ is rather large.

From this analysis we conclude that it is not possible to explain both the 3-jet distribution and the total cross section with an abelian vector gluon theory. In the $\overline{MS}$ scheme R can be adjusted, but not the 3-jet cross section, whereas in a low-order scheme the jet distribution can be described, but not R.

After this detour we come back to QCD. So far, the comparison of the higher order corrections with data was based on the cluster analysis. This is not the only possibility. Another way is to take 3- and 4-jet contributions together. An example for such 3-4 jet inclusive distributions is the acollinearity distribution also called energy-energy correlation, defined in Sect.3.1.5. In higher order QCD this includes the correlations originating from 3- and 4-parton production which depend just on one angle χ. From the energy-energy correlation one can derive the asymmetry cross section defined in (3.1.43) in Sect.3.1.5. This asymmetry cross section is not sensitive to inclusion of 2-jet contributions, which are present in the energy-energy correlation already in lowest order. As an example, we show the asymmetry cross section as a function of $\cos\chi$ in Fig.3.47. It includes 3-jet contributions of order α_s and α_s^2 and the 4-jet terms. The curve is calculated for $\varepsilon = 0.2$ and $\delta = 30°$ with $\alpha_s = 0.14$. The theoretical prediction /Schneider, Kramer and Schierholz, 1983/ is compared to the CELLO data shown earlier /Behrends et al., 1982b/ (see Fig.3.36). These data are corrected for radiative and other effects but contain still the effects of fragmentation. We estimate the influence of fragmentation to increase α_s approximately by 0.01 for the

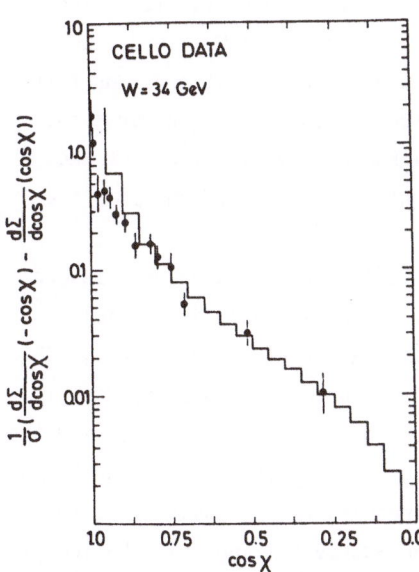

CELLO DATA

W = 34 GeV

Fig.3.47. Asymmetry cross section AS($\cos\chi$) as predicted by first and second order QCD with $(\varepsilon,\delta) = (0.2, 30°)$ and $\alpha_s = 0.14$ compared to CELLO data at W = 34 GeV. These data contain still fragmentation effects

independent fragmentation models of Ali and Hoyer. Therefore we quote from this fit $\alpha_s = 0.14 \pm 0.02$.

In order to take the fragmentation of quarks and gluons into account, one starts, as in low-order, from the formula (3.2.10) for the 3-jet cross section $d^2\sigma/dx_1 dx_2$. This formula includes the $O(\alpha_s^2)$ corrections, and it is used to calculate the probability for the production of 3 jets. From here on one proceeds as it was done in low-order described in detail in Sect.2.4. Equation (3.2.10) includes the contribution of 4 partons inside the resolution criteria given by (ϵ, δ) or y parameters. Of course, the 4-parton contributions outside the ϵ, δ or y cuts, which produce genuine 4 jets and which are not contained in (3.2.10), must be added explicitly. In these contributions q, $\bar{q}$ and the two g's (in $q\bar{q}gg$) and two q's and $\bar{q}$'s (in $q\bar{q}q\bar{q}$), respectively, are supposed to fragment independently into hadrons. This way all hadron distributions of interest can be calculated. The parameters which are more or less free in the fragmentation model, like σ_q, a, r etc. introduced in Sect.2.4, are adjusted simultaneously with α_s to the data. Such a model based on the 3-jet formulae (3.2.10) with $\epsilon = 0.2$ and $\delta = 40°$ has been developed by members of the TASSO Collaboration /Wu, 1983/ and compared to their experimental data. They used for fragmentation the Hoyer model with fragmentation of 4 jets added. In this model 3 (or 4) partons fragment independently according to the Field-Feynman model with quark and gluon fragmentation assumed to be equal (see Sect.2.4). We show three examples of their curves. In Fig.3.48 the sphericity distribution is plotted and in Figs.3.49,50 we see the planarity distribution where planarity is defined as $P = Q_2 - Q_1$ (see Sect.2.3 for the definition of the Q_i) and the $<p_{T\ out}^2>$ distribution, where $p_{T\ out}$ is the average transverse momentum out of the event plane. The resulting second-order [first-order] α_s is 0.168 ± 0.003 (stat.) ± 0.03 (syst.) [0.194 ± 0.005 (stat.) ± 0.03 (syst.)]. The parameters of the fragmentation model can be found in the paper of Wu /1983/. We see that the event shape distributions are well described by either first order or first and second order theory. The resulting value of α_s are not the same, the first and second order result being 13% lower. These results agree nicely with those of the JADE Collaboration /Bartel et al., 1982a/ although the two methods of analysis are quite different.

The test of the theory as performed by the TASSO Collaboration based on measured hadron distribution is certainly an alternative to the cluster analysis. A similar analysis has been done also by the MARK J Collaboration at PETRA /Adeva et al., 1983c/. They fitted the first and second order theory to the measured acollinearity distribution and the asymmetry correlation. For distinguishing 2, 3 or 4 jets they used the Sterman-Weinberg parameters $\epsilon = 0.15$ and $\delta = 26°$. They have performed many checks on the cut-off parameters, ϵ and δ, and found that the results are insensitive to the variation of these parameters in the range $\epsilon = 0.15$-0.30. By fitting only the $|\cos\chi|$

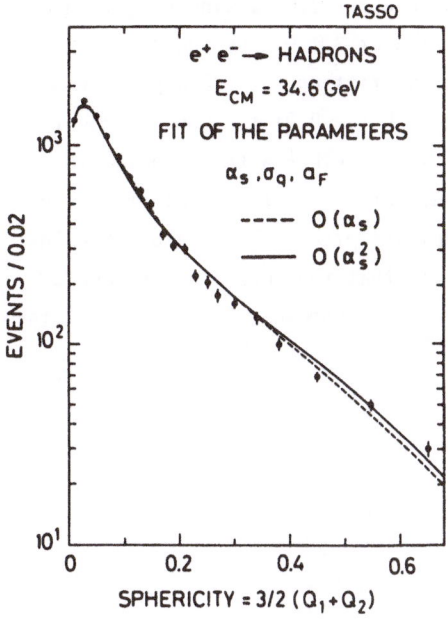

Fig.3.48

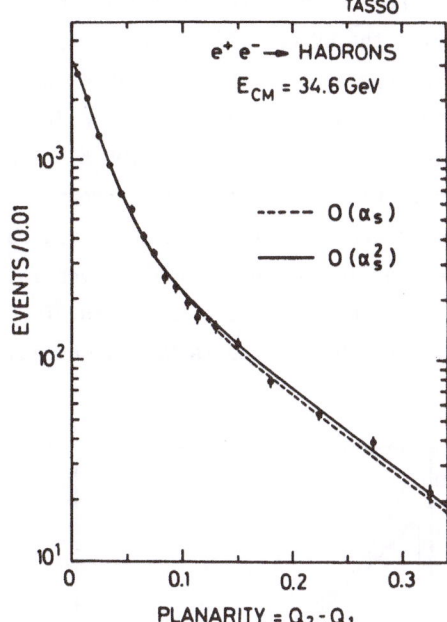

Fig.3.49

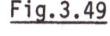

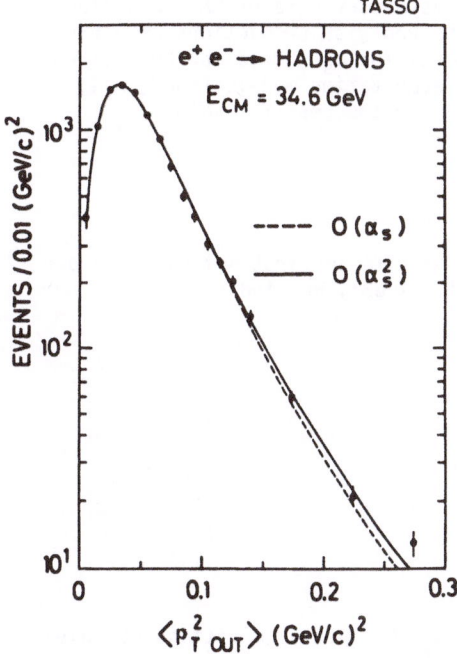

Fig.3.50

Fig.3.48. Sphericity distribution from TASSO Collaboration. Dashed and solid line are respectively results of simultaneous fits in first and first + second order QCD with Hoyer model fragmentation and $(\varepsilon, \delta) = (0.2, 40°)$ for 2-, 3- and 4-jet separation

Fig.3.49. Same as Fig.3.48 for planarity distribution

Fig.3.50. Same as Fig.3.48 for $\langle p^2_{T\ out} \rangle$ distribution

< 0.72 region they are explicitly insensitive to the δ cut over a wide range. In contrast to the CELLO Collaboration study mentioned in Sect.3.1.9 they found also very little dependence on the fragmentation model used. By fitting the QCD asymmetry prediction to the data for $|\cos\chi| < 0.72$ they find $\alpha_s = 0.14 \pm 0.01$ for Lund fragmentation and $\alpha_s = 0.12 \pm 0.01$ for the Ali model. This fit, together with the data, is shown in Fig.3.51. The curve without fragmentation, but $\alpha_s = 0.13$, is plotted also. The data are not corrected for photon radiation and acceptance. These are taken into account in the theoretical curves instead. We see that the second order corrections reduce the value of α_s again. The result for α_s is somewhat lower than obtained by the JADE and the TASSO Collaboration.

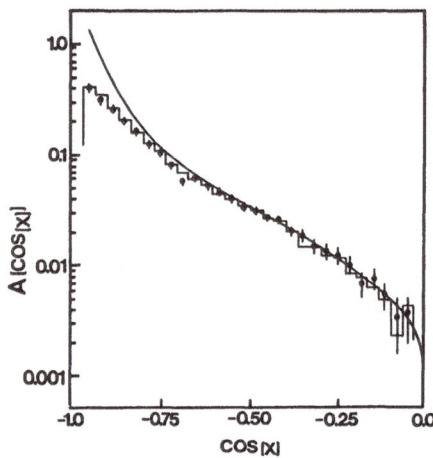

Fig.3.51. Asymmetry data of MARK J Collaboration compared with predictions at parton level (curve) for $\alpha_s = 0.13$ and predictions for two fragmentation models (Lund, Ali; histogram) for the best fit values α_s

Table 3.3. Strong coupling constant α_s obtained from fits in first and second order QCD. References: Bartel et al. /1982/, Adeva et al. /1983/, Wu /1983/

Collaboration	α_s
CELLO	$0.14 \pm 0.02 \pm 0.03$
JADE	$0.16 \pm 0.015 \pm 0.03$
MARK J	$0.13 \pm 0.01 \pm 0.02$
TASSO	$0.168 \pm 0.003 \pm 0.03$

In Table 3.3 we have collected the results for α_s obtained from fitting QCD distributions up to α_s^2. The average value is $\alpha_s = 0.15 \pm 0.015$. This can be converted into a value of $\Lambda(\overline{MS})$ using the second order formula (3.2.38) for Λ ($N_f = 5$). The result is $\Lambda_{\overline{MS}} = (0.30 \pm 0.17)$ GeV. This is in very good agreement with the Λ values obtained recently from analysis of deep inelastic lepton-nucleon scattering data. For example, Abramowicz et al. /1983/ obtained from the analysis of their neutrino-scattering data (CDHS Collaboration) $\Lambda_{\overline{MS}} = (0.25 \, {}^{+0.15}_{-0.10})$ GeV. Deveto, Duke, Owens and Roberts /1983/ report from a combined analysis of nucleon structure functions obtained in neutrino, electron and muon scattering experiments $\Lambda_{\overline{MS}} = (0.35 \pm 0.10)$ GeV whereas Barker, Martin and Shaw /1983/ conclude from an analysis of all available neutrino xF_3 data $\Lambda_{\overline{MS}} = (0.30 \pm 0.13)$ GeV.

In this approach we must ask the question to what extent the values we have obtained for α_s are really independent of the choice of the parameters ε, δ or y, respectively, for separating 3 and 4 jets or is there more or less a unique choice for these parameters. Let us discuss this problem in connection with the y parameter first.

We remarked already that in the cluster analysis this problem does not arise, since the separation parameter y in the evaluation of the perturbation theory has its equivalent in the cluster analysis. In the discussion of $O(\alpha_s)$ perturbation theory we noticed already that exclusive jet cross sections, like 2- and 3-jet cross sections, have reasonable values only, if y lies in the vicinity of 0.05. This applies equally well to the jet distributions $d\sigma^{3-jet}(y)/dT$ and $d\sigma^{4-jet}(y)/dT$. Also in these cross sections the expansion parameter is essentially $C_F(\alpha_s/\pi)\ln^2 y$ [see for example (3.2.14-16)] as in $\sigma^{2-jet}(y)$. Therefore $C_F(\alpha_s/\pi)\ln^2 y$ should not be too large which bounds y to ≥ 0.02. In the inclusive distribution $d\sigma^{3-jet}(y)/dT + d\sigma^{4-jet}(y)/dT$, in which 3 and 4 jets are not distinguished anymore, the "large" logarithmic terms proportional to $(\alpha_s/\pi)\ln^2 y$ and $(\alpha_s/\pi)\ln y$ cancel (in agreement with the Kinoshita-Lee-Nauenberg theorem) and we would expect that the inclusive jet distribution is independent of y. Unfortunately this is not the case. In order to show this we define for the inclusive 3- and 4-jet distribution $(1/\sigma^{(2)})d\sigma/dT$ the function $A_1(T)$ which is the $O(\alpha_s^2)$ contribution of this distribution as a function of T /Gutbrod, Kramer and Schierholz, 1983; Kramer, 1982,1983/:

$$\frac{1}{\sigma^{(2)}} \frac{d\sigma}{dT} = \frac{\alpha_s}{\pi} A_0(T) + \left(\frac{\alpha_s}{\pi}\right)^2 A_1(T) \quad . \tag{3.2.21}$$

In Fig.3.52 $A_1(T)$ is represented for two y values 0.04 and 0.001 together with $A_0(T)$ as a function of T. $A_1(T)$ increases by almost a factor of 2 if y is decreased by a factor of 40. Therefore, for $y = 0.04$ the $O(\alpha_s^2)$ correction of the *inclusive* thrust distribution is approximately 50% whereas for $y = 0.001$ it is approximately 90% ($\alpha_s/\pi = 0.05$). Thus, we must conclude that $A_1(T)$ depends still on y, and it is essential for which y the inclusive thrust distribution will be used in the analysis. The dependence of $A_1(T)$ on y (in Fig.3.53 it is y^{-1}) for some T intervals: $0.900 \leq T \leq 0.915$ etc. is shown in Fig.3.53 together with asymptotic values for $y = 0$ obtain-

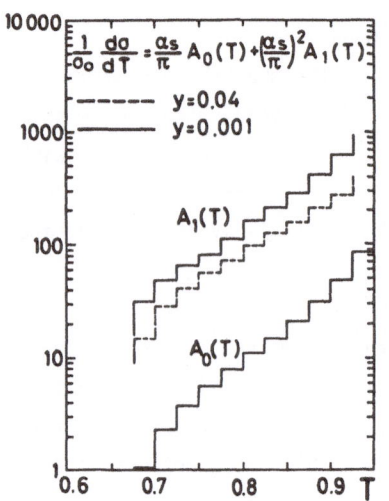

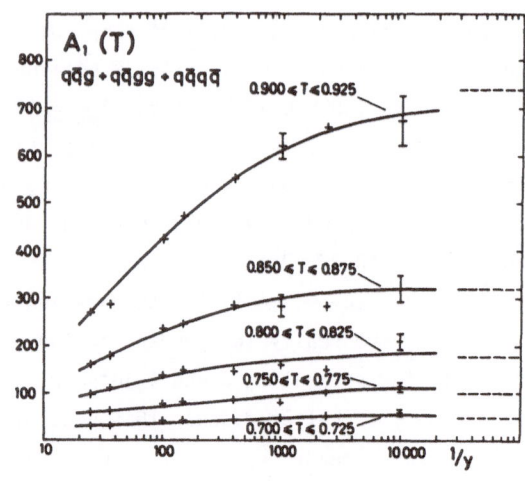

Fig.3.52. The sum of three-jet $[O(\alpha_s^2)]$ and four-jet contribution to $A_1(T)$ as a function of T for y = 0.04 and y = 0.001 together with lowest order contribution $A_0(T)$

Fig.3.53. $A_1(T)$ for different thrust bins as a function of 1/y. The dashed lines are the asymptotic values of $A_1(T)$ obtained by ELLIS and Ross /1981/

ed by Ellis and Ross /1981/. We see that $A_1(T)$ approaches this asymptotic value already at $y = 10^{-4}$. The results, shown in Fig.3.53, have computational errors caused by the Monte-Carlo integration of the 4-parton contributions outside the y cut-off. For $y = 10^{-4}$ we have given this error explicitly which results from a 1% error in the Monte-Carlo integration. This error is increased at small y since $A_1(T)$ is computed by adding the large $A_1(T)_{4-jet}$ with the correspondingly large negative $A_1(T)_{3-jet}$.

A similar study was made for the ε, δ case /Gutbrod, Kramer and Schierholz, 1983; Kramer, 1982,1983/. To reduce the problem to a one-parameter problem $\varepsilon = (1-\cos\delta)/2$ was assumed, so that with decreasing ε also δ becomes smaller. The results are shown

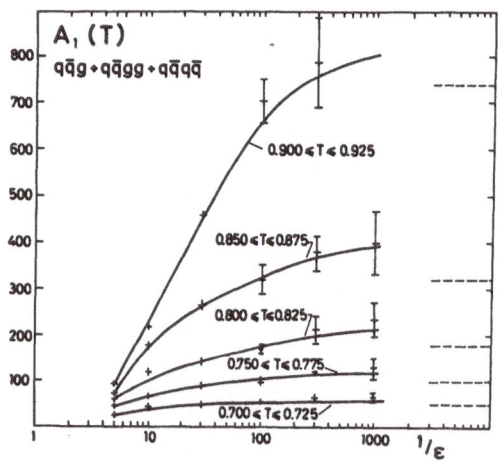

Fig.3.54. $A_1(T)$ for different thrust bins as a function of $1/\varepsilon$ [$\varepsilon = (1-\cos\delta)/2$]

in Fig.3.54. For $\varepsilon \to 0$ the function $A_1(T)$ approaches the same asymptotic values as in Fig.3.53 for $y \to 0$, taking into account that the Monte-Carlo errors are somewhat larger now. This comparison of the asymptotic $A_1(T)$ for $y \to 0$ or $\varepsilon, \delta \to 0$, respectively, could certainly be improved by applying better numerical methods. But since the asymptotic value is not of particular interest for the data analysis this has not been done. The asymptotic values for the energy correlation and the asymmetry cross section have been computed by Ali and Barreiro /1982/ and by Richards, Stirling and Ellis /1982/.

The increase of $A_1(T)$ with decreasing y has its origin in the fact that $A_1(T)$ is the sum of two distributions in which the variable thrust is defined differently. For fixed y the thrust T in $d\sigma^{3-jet}(y,T)/dT$ is calculated according to $T = x_{max} = \max(x_1, x_2, x_3)$ where the x_i are the energies of the 3 jets, and in $d\sigma^{4-jet}(y,T)/dT$ it is calculated from the momenta of the 4 parton final state. This means that in the 3-jet part T is calculated from other configurations than in the 4-jet part. How the formula for T depends on the number of particles in the final state can be seen in (2.3.4) of Sect.2.3. Of course, this would not cause any problem if experimentally 3 and 4 jets can be separated from each other so that the thrust distribution $(1/\sigma)d\sigma/dT$ for both classes of events can be measured and added. This route is taken in the cluster analysis, and we have seen that the parameter y for separating 3 and 4 jets has its correspondence in the cluster analysis. This means, in the cluster analysis $A_1(T)$ is measured (after subtracting the $A_0(T)$ part) for a particular y, which is compared with the theoretical predictions for the same y. Concerning the α_s^2 corrections these two parts, 3 jets and 4 jets, have for $y = 0.04$ and $y = 0.01$ the structure as shown in Fig.3.55. For $y = 0.04$ the $A_1(T)$ is mostly 3 jets and much less

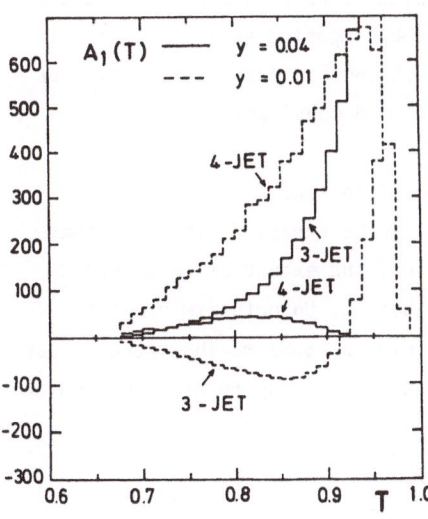

Fig.3.55. Three- and four-jet components of $A_1(T)$ for $y = 0.04$ and $y = 0.01$ as a function of T

4 jets whereas at $y = 0.01$ the 3 jets give already a negative contribution and $A_1(T)$ is mostly 4 jets.

The sum of these two contributions changes with y as seen in Fig.3.53. This should also be borne out in the experimental data. But it has not been tested yet. Thus, if y is changed, the inclusive distribution, i.e. $A_1(T)$ changes, since part of the final states, which for larger y contribute to $d\sigma^{3-jet}(y,T)$, are shifted to $d\sigma^{4-jet}(y,T)$ when y is decreased. Since the T values for these two event classes, 3 and 4 jet, differ the sum changes. It seems that terms proportional to $y\ln y$ with relatively large coefficients are changed in $d\sigma^{4-jet}(y,T)$ when 4 partons are combined to 3 jets with subsequent calculation of T. Suppose, we have chosen $y = 0$ for the calculation of $A_1(T)$, then we have included the maximal amount of 4 jets, no matter how soft an individual parton may be or how collinear a parton may be. For this situation the notion "bare jets" and "bare thrust" was introduced by Gottschalk /1982/ in contrast to "dressed jets" and "dressed thrust" for the case of a non-zero y. Thus, Figs. 3.53,54 tell us how $A_1(T)$ changes if the resolution of dressed jets is varied and how it approaches the limit for bare jets. This limit or the bare thrust distribution has been calculated by Ellis and Ross /1981/, Kunzst /1980,1981/ and Ali /1981/ based on the work of Ellis, Ross and Terrano /1980,1981/ and independently by Vermaseren, Gaemers and Oldham /1981/. Originally, it was thought that the results of Fabricius, Schmitt, Schierholz and Kramer /1980/ and those of Ellis, Terrano and Ross /1980,1981/ and Vermaseren, Gaemers and Oldham /1981/ are in apparent conflict with each other. This controversy is now solved. Ellis, Ross and Terrano and other authors, who based their calculations on this work, evaluated the jet variable distributions for "bare jets" whereas Fabricius, Kramer, Schierholz and Schmitt calculated the shape distributions for "dressed jets". Later Gottschalk /1982/ has shown on the basis of an abelian gluon model that the results of Fabricius et al. could be recovered from the theory of Ellis, Ross and Terrano by imposing an additional jet resolution criterion à la Fabricius et al. Figures 3.53,54 now demonstrate this for the full QCD by starting with a finite jet resolution and considering its approach to the limit of bare jets.

This dependence of inclusive distributions (sum of 3 jet and 4 jet) on the parameters that define the resolution of 3 and 4 jets is to be expected also for other distributions and not only for the thrust distribution. The extent of this resolution dependence may vary from distribution to distribution. Unfortunately, these effects have not been studied yet in great detail. In Figs.3,56,57 we show results for two other quantities, the energy correlation function (3.1.35) up to order α_s^2, which is expressed in the form [in analogy to (3.2.21)]

$$\frac{1}{\sigma^{(2)}} \frac{d\Sigma}{d\cos\chi} = \frac{\alpha_s}{\pi} C(\cos\chi) + \left(\frac{\alpha_s}{\pi}\right)^2 D(\cos\chi) \qquad (3.2.22)$$

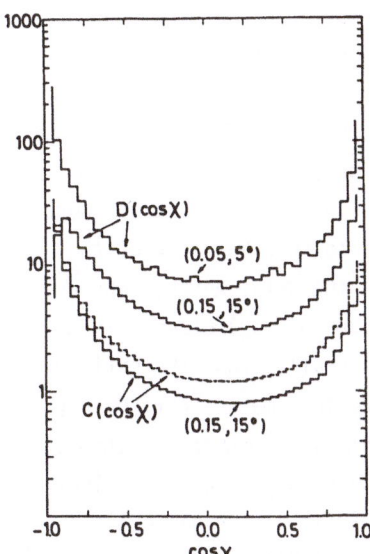 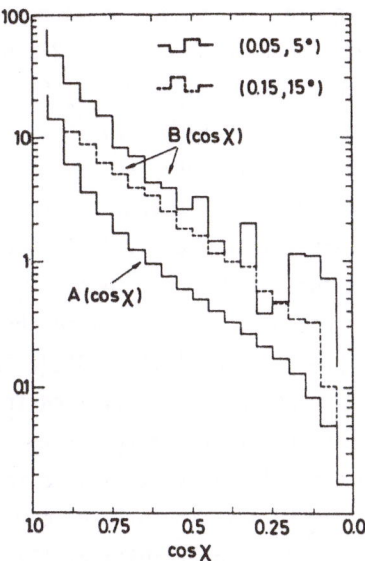

Fig.3.56. First [C(cosχ)] and second order [D(cosχ)] QCD prediction for energy correlation cross section as a function of cosχ for (ε,δ) = (0.15, 15°) and (ε,δ) =
(0.05, 5°)

Fig.3.57. First [A(cosχ)] and second order [B(cosχ)] QCD prediction for asymmetry
cross section as a function of cosχ for (ε,δ) = (0.15, 15°) and (ε,δ) = (0.05, 5°)

and the asymmetry distribution [see (3.1.43) for the definition] which is written
as /Schneider, Kramer and Schierholz, 1983/:

$$AS(\cos\chi) = \frac{\alpha_s}{\pi} A(\cos\chi) + \left(\frac{\alpha_s}{\pi}\right)^2 B(\cos\chi) \quad . \tag{3.2.23}$$

The higher order contributions D(cosχ) and B(cosχ) have been calculated for two cases
of (ε,δ) values: a large one (ε,δ) = (0.15, 15°) and a rather small one (ε,δ) = (0.05,
5°). D and B contain the 3- and 4-jet contribution to the energy correlation and the
asymmetry correlation respectively. We see from Fig.3.56 that D(cosχ) changes by more
than a factor of two when ε and δ are varied in this rather small region. This occurs
for all angles χ between 0 and π. The asymptotic value of D(cosχ) for ε,δ → 0 is even
larger, approximately a factor of two above the (ε,δ) = (0.05, 5°) curve, if we compare with the results of Ali and Barreiro /1982/ who calculated the limit of D(cosχ)
on the basis of the Ellis, Ross and Terrano approach. Therefore the α_s^2 terms in the
energy correlations show the same resolution dependence as the inclusive thrust distribution. In this comparison we must take into account, however, that the lowest order contribution, i.e. C(cosχ), is already resolution parameter dependent. As we remarked earlier, the energy correlation function, as defined in Sect.3.1.5, contains
also the contributions corresponding to 2 jets in the qq̄g final state. If this contri-

bution is subtracted the cross section is reduced. The amount of reduction depends on the ε,δ parameters chosen to separate 2 and 3 jets. In Fig.3.56 we show $C(\cos\chi)$ as a function of $\cos\chi$ for $(\varepsilon,\delta) = (0.15, 15°)$ (lower curve) and with the 2-jet region included, i.e. $\varepsilon,\delta = 0$ (upper curve). This latter curve is equal to the cross section in Fig.3.9 except for the factor α_s/π. As can be seen the omission of the 2-jet contribution changes the energy correlation function already by 40%. Of course, when comparing to experimental data, it is very important to know how much of the 2-jet contribution contained in $e^+e^- \rightarrow q\bar{q}g$ is included in the data. Another observation is that the ratio of the α_s^2 term to the α_s term, i.e. $D(\cos\chi)/C(\cos\chi)$ is only of the order of 4 [for $(\varepsilon,\delta) = (0.15, 15°)$]. This is much smaller than the ratio of $A_1(T)/A_0(T)$ for $y = 0.04$, which is obtained from Fig.3.52. This means that the higher order corrections in the energy-energy correlations are smaller than in the thrust distribution. The situation is very similar for the asymmetry cross section. $B(\cos\chi)$ changes approximately by a factor of two between $(\varepsilon,\delta) = (0.15, 15°)$ and $(\varepsilon,\delta) = (0.05, 5°)$. If compared to the results of Ali and Barreiro /1982/ the curve for $(\varepsilon,\delta) = (0.05, 5°)$ gives already the results obtained in the limit $(\varepsilon,\delta) \rightarrow 0$. So in the asymmetry cross section the asymptotic value is reached already rather early. Furthermore the ratio B/A is only near four, so that the effect of higher order corrections is reduced here also. Thus it seems that for the energy-energy correlations and the asymmetry cross section the convergence of the perturbative series is better as in the thrust distribution.

The only problem which still has to be solved is the choice of the parameters ε, δ and y, respectively, in such model calculations, in which fragmentation is added to the primordial production of quarks and gluons. The same problem arose in connection with the fragmentation of 2 and 3 jets based on $O(\alpha_s)$ perturbation theory. Obviously the parameters ε, δ and y for separating 3 and 4 jets should be of the same order as the parameters used for separating 2 and 3 jets in $O(\alpha_s)$, i.e. $\varepsilon \simeq 0.2$, $\delta \simeq 40°$ and $y \simeq 0.05$, respectively. We have seen already that $y = 0.05$, for example, yields reasonable results for jet multiplicities. Furthermore it would not make much sense to choose y much smaller than the non-perturbative jet spread caused by the fragmentation. At $W = 30$ GeV this width lies between 0.05 and 0.1, as the model result for 2 jets in Fig.3.4 indicates ($y = 1-T$). Experimentally the width is very well represented by the distribution of m_1^2/W^2, the mass of the narrow jet shown in Fig.3.58, which also yields $\Delta(m_1^2/W^2) \simeq 0.05$. Of course, these considerations do not fix these parameters uniquely. They give us just the order of magnitude. Therefore, in the fragmentation models ε, δ or y must be considered more or less as free parameters, which must be determined simultaneously with the other free parameters of the model through the fit to experimental data. Then, the fragmentation parameters are supposed to change with varying y. A change of the fragmentation parameters occurs already when

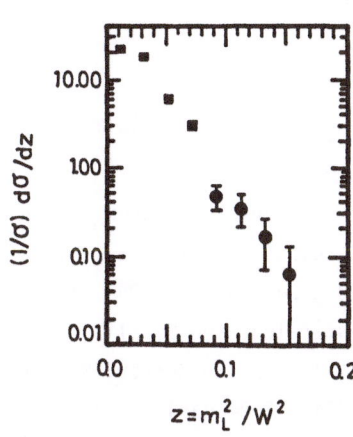

Fig.3.58. JADE data about narrow jet mass (m_1^2/W^2) distribution

$z = m_L^2/W^2$

comparing first and second order fits /Wu, 1983/. We also note that for $T \leq 0.82$ the variation of $A_1(T)$ with decreasing y or ε is moderate, if y (or ε^2 and δ^2) are changed by less than a factor of 2 as can be seen in Figs.3.53,54. Therefore the influence on determinations of α_s is not so large. Of course, it is desirable to have physical observables which depend only very little on these resolution parameters.

3.2.3 Evidence for Four-Jet Production

As remarked in the previous section, in order α_s^2 final states with four partons, $e^+e^- \to q\bar{q}gg$ and $e^+e^- \to q\bar{q}q\bar{q}$, can also be produced, which should appear at high enough center-of-mass energies W and under the right kinematical conditions as four clearly resolved jets. Their production rate is expected to be reduced by a factor of α_s compared to the 3-jet rate. The discovery of these 4-jet events with properties as predicted by QCD perturbation theory may serve as further evidence for the realization of QCD in nature. Studying 4-parton final states can also be useful to distinguish abelian and non-abelian vector gluon theories, as is evident from the existence of additional diagrams with the 3-gluon coupling in the case of QCD (see Fig.3.37). We shall see later that it is not a simple matter to verify the full gauge structure, i.e. the gluon self-interaction, in 4 jets alone.

The 4-jet cross section is determined from the two groups of diagrams in Fig.3.37 which lead to the final states

$$e^+(p_+) + e^-(p_-) \to q(p_1) + \bar{q}(p_2) + g(p_3) + g(p_4) \tag{3.2.24}$$

and

$$e^+(p_+) + e^-(p_-) \to q(p_1) + \bar{q}(p_2) + q(p_3) + \bar{q}(p_4) \quad . \tag{3.2.25}$$

The differential cross section is calculated from

$$d\sigma = \frac{e^4}{(2\pi)^8 \, 2q^6 N_s} \, \{p_+, p_-\}_{\mu\nu} \, \prod_{i=1}^{4} \frac{d^3 p_i}{2p_{i0}} \, \delta^{(4)}\left(p_+ + p_- - \sum_{i=1}^{4} p_i\right) H^{\mu\nu} \tag{3.2.26}$$

with $q = p_+ + p_-$. $\{p_+, p_-\}_{\mu\nu}$ is the well-known lepton tensor. The hadron tensor $H_{\mu\nu}$ contains the complete information about the final states (3.2.24,25). N_s is a statistical factor depending on the number of identical particles in (3.2.24) and (3.2.25). The complete formulae for $e^+e^- \to q\bar{q}gg$ and $e^+e^- \to q\bar{q}q\bar{q}$ were first calculated by Ali et al. /1979,1980/. They are explicitly written down in Ali et al. /1980/, in Körner, Schierholz and Willrodt /1981/ and in Ellis, Ross and Terrano /1981/. Some early studies about $e^+e^- \to q\bar{q}q\bar{q}$ can be found in De Grand, Ng and Tye /1977/.

The differential cross section (3.2.26) depends on 5 independent variables which describe the position of the four jets in phase space. They can be chosen as

$$x_i = \frac{2|\mathbf{p}_i|}{W} \, , \quad i = 1,2,3 \quad \text{and} \quad x_{ij} = \frac{2|\mathbf{p}_i + \mathbf{p}_j|}{W} \, , \quad i,j = 12,13 \quad . \tag{3.2.27}$$

In addition $d\sigma$ depends on two angles θ and χ, which determine the orientation of the 4-jet event with respect to the e^- beam direction similar as in the 3-jet case. Details about this dependence can be found in the work of Ali et al. /1980/ and Clavelli and von Gehlen /1983/ where also the effects of Z exchange are considered.

In principle the hadron tensor $H_{\mu\nu}$ has a very rich structure which can be expressed in various multi-dimensional distributions and angular correlations. Because of the limited statistics only one-dimensional distributions, integrated over angles θ and χ, have been studied experimentally. For such a distribution acoplanarity is a very useful variable since 4-jet events stand out against 2- and 3-jet events by having a non-vanishing acoplanarity A. Thus $d\sigma/dA$ is the canonical quantity to analyse, as it allows one to cut-off the dominant 2- and 3-jet events experimentally. Variables similar to A like the aplanarity Ap (see Sect.2.3) can be chosen as well.

Distributions like $d\sigma/dT$ are less useful, since they receive contributions also from the infrared singular region of phase space. Therefore the thrust distribution can be defined only, if the singular regions are cut-off by boundaries in the invariant mass, i.e. by y, or by ε, δ cuts. Such distributions can be compared only with events which have been identified in the cluster analysis as 4-jet events. Otherwise we can consider the thrust distribution for a superposition of 3- and 4-jet events as discussed in the previous section, for which, as we know, the 4-jet contribution is rather small (say, for y = 0.04). In this sense it is rather unique to study the acoplanarity distribution first. The distribution $(1/\sigma^{(2)})d\sigma/dA$ is represented in Fig.3.59, normalized to the zeroth-order cross section $\sigma^{(2)}$ for W = 40 GeV ($\alpha_s = 0.20$)

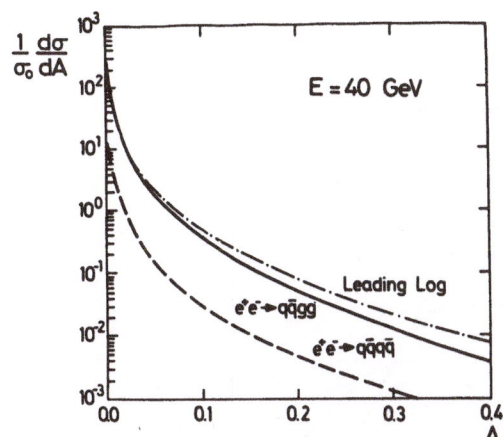

$\frac{1}{\sigma_0}\frac{d\sigma}{dA}$

E = 40 GeV

Leading Log

$e^+e^- \rightarrow q\bar{q}gg$

$e^+e^- \rightarrow q\bar{q}q\bar{q}$

A

Fig.3.59. Acoplanarity distributions $(1/\sigma)d\sigma/dA$ for $e^+e^- \rightarrow q\bar{q}gg$ (full curve) and $e^+e^- \rightarrow q\bar{q}q\bar{q}$ (dashed curve) together with leading-log approximation (dashed-dotted curve) for $e^+e^- \rightarrow q\bar{q}gg$

and for $e^+e^- \rightarrow q\bar{q}gg$ and $e^+e^- \rightarrow q\bar{q}q\bar{q}$ separately. The total distribution is obtained from the sum of these two components.

The differential cross section $d\sigma/dA$ diverges for $A \rightarrow 0$. For $e^+e^- \rightarrow q\bar{q}gg$ the leading logarithm behaviour is:

$$\frac{1}{\sigma^{(2)}}\frac{d\sigma}{dA} = \frac{1}{2}\, C_F^2\Big(\frac{\alpha_s}{\pi}\Big)^2 \frac{1}{A}\,|\ln A|^3 \qquad (3.2.28)$$

as $A \rightarrow 0$. The leading log formula can be seen to give a good description of the differential A distribution up to rather large A values. The equivalent formula for $e^+e^- \rightarrow q\bar{q}q\bar{q}$ is:

$$\frac{1}{\sigma^{(2)}}\frac{d\sigma}{dA} = \frac{1}{16}\, C_F^2 N_f\Big(\frac{\alpha_s}{\pi}\Big)^2 \frac{1}{A}\,|\ln A|^2 \quad . \qquad (3.2.29)$$

So the leading log behaviour is one power in $|\ln A|$ less than for $e^+e^- \rightarrow q\bar{q}gg$. The contribution of $e^+e^- \rightarrow q\bar{q}gg$ dominates, $e^+e^- \rightarrow q\bar{q}q\bar{q}$ is 10 times smaller over most of the region of A. For four massless final state particles A is bounded between 0 and 2/3. The maximal A value occurs for the configuration, where the four momenta point from the center of a tetrahadron [with side length $(q^2/6)^{1/2}$] to its four corners which gives A = 2/3 according to (2.3.6). We see that $d\sigma/dA$ decreases strongly with increasing A so that events with large A are rare.

Because of the singularity for $A \rightarrow 0$ [see (3.2.28,29)] the differential cross section is not integrable over A. As we discussed in Sect.3.2.1, this is to be expected because of the infrared singularities associated with emission of soft and collinear quarks, antiquarks and gluons (3.2.24,25). These singularities cancel against the corresponding one- and two-loop virtual corrections in Figs.3.38,39 so that the total cross section to order α_s^2 becomes finite. Due to the singular behaviour of $d\sigma/dA$

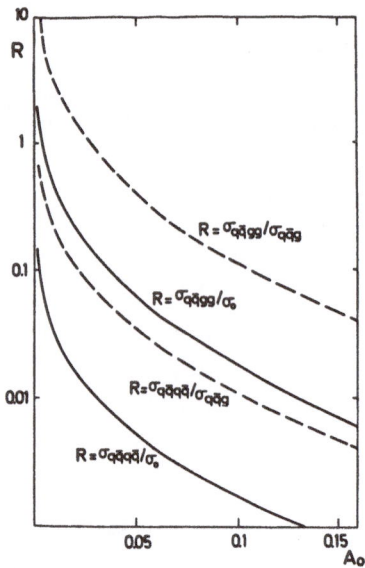

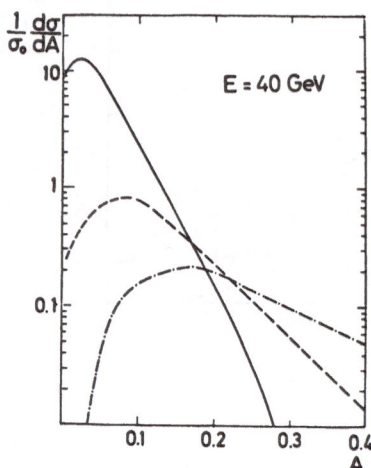

Fig.3.60. Integrated four-jet cross section as a function of acoplanarity cut-off A_0 for $e^+e^- \to q\bar{q}gg$ and $e^+e^- \to q\bar{q}q\bar{q}$ separately. Ratio $\sigma(4\text{-jet})$ to $\sigma(3\text{-jet})$ as a function of A_0 for T cut-off $T_0 = 0.9$ in $\sigma(3\text{-jet})$

Fig.3.61. Acoplanarity distribution $(1/\sigma)d\sigma/dA$ for $e^+e^- \to q\bar{q}$ (full curve), $e^+e^- \to q\bar{q}g$ (dashed curve) and $e^+e^- \to q\bar{q}gg + q\bar{q}q\bar{q}$ (dashed-dotted curve) with fragmentation according to the Ali model

as $A \to 0$ the differential distribution should be considered to be reliable only for values of A above some cut-off A_0. Integrating $d\sigma/dA$ from 2/3 to A_0 one obtains a cut-off dependent 4-jet cross section $\sigma(A_0)$, shown in Fig.3.60 for $e^+e^- \to q\bar{q}gg$ and $e^+e^- \to q\bar{q}q\bar{q}$ separately. The ratio to $\sigma^{(2)}$ measures the 4-jet multiplicity in these two channels. A cut-off value for which $\sigma(A_0)/\sigma^{(2)} \simeq \alpha_s^2 \simeq 0.04$ should be considered a reasonable choice, above which a perturbatively calculated $d\sigma/dA$ can be trusted. From Fig.3.61 this corresponds to $A_0 = 0.07$. Of course, these considerations are analogous to the 3-jet case and the cut-off's y and T_0. We also show in Fig.3.60 the ratio of 4-jet to 3-jet production as a function of A_0. The 3-jet cross section is computed in $O(\alpha_s)$ with a thrust cut-off at $T_0 = 0.9$. We see that this ratio is quite large if the A cut-off is chosen in the vicinity of 0.1.

One of the aims for studying 4 jets is to see how large the influence of the three-gluon coupling is. In some sense this is an ill-defined question since, on the one hand, the relative contribution of the three-gluon coupling depends on the gauge choice and on the other hand theories with only global SU(3) symmetry are not renormalizable (although for tree-diagrams considered in this section renormalization is not relevant). Therefore we compare to the abelian gluon theory with the number of quark flavours multiplied by 3, already considered in the previous section and denoted QAD. It turns out that most of the distributions normalized to $\sigma^{(2)}$ remain more or less unchanged if the QAD coupling $\alpha_A = 4\alpha_s/3$, i.e. if α_A is adjusted to the same 3-jet cross section as in QCD /Gaemers and Vermaseren, 1980; Nachtmann and Reiter, 1982a/. However, the partition of the cross section for the two final states (3.2.24) and (3.2.25) is different in QAD. For $A_0 = 0.05$ and the same zeroth-order total cross section $\sigma^{(2)}$ we have

$$\sigma(\text{QAD, } q\bar{q}gg) \simeq 0.8 \ \sigma(\text{QCD, } q\bar{q}gg)$$

$$\sigma(\text{QAD, } q\bar{q}q\bar{q}) \simeq 8 \ \sigma(\text{QCD, } q\bar{q}q\bar{q})$$

$$(3.2.30)$$

so that in the abelian theory the final states $q\bar{q}gg$ and $q\bar{q}q\bar{q}$ are produced roughly with equal probability. In QCD this ratio is approximately 10:1. In case one would be able to distinguish quark and gluon jets the dominance of the $q\bar{q}gg$ channel could be established. Unfortunately this is still very difficult (see however Bartel et al. /1983/), so that QAD could not be excluded by 4-jet studies alone up to now. If both channels are added we have

$$\sigma(\text{QAD, } q\bar{q}gg + q\bar{q}q\bar{q}) \simeq 1.3 \ \sigma(\text{QCD, } q\bar{q}gg + q\bar{q}q\bar{q})$$

$$(3.2.31)$$

which means that the integrated 4-jet cross sections are roughly equal in QAD and QCD if the same cut-off A_0 is applied and $\alpha_A = 4\alpha_s/3$. We made use of this fact in the previous section by stating that in QCD and in QAD the 4-jet rates are equal, if α_A is adjusted to the low-order 3-jet cross section.

As our next point we study, how $d\sigma/dA$ is changed by the fragmentation of quarks and gluons into hadrons, and how large the background is, originating from $q\bar{q}$ and $q\bar{q}g$ final states. The comparison with $d\sigma/dA$ for non-perturbative 2-jet production calculated in a Field-Feynman model is presented in Fig.3.61. This distribution which includes also weak decay effects of c and b quarks /Ali, Körner, Kramer and Willrodt, 1980b/ is still rather broad at W = 40 GeV. The input p_T for the Field-Feynman model was chosen as $\sigma_q = \langle p_T^2 \rangle^{1/2} = 0.25$ GeV. (Actually, according to our present knowledge σ_q should be larger, $\sigma_q \simeq 0.3$ GeV, which makes the distribution even broader.) The average A for this non-perturbative 2-jet distribution is calculated to be $\langle A \rangle_{\text{non-pert.}} = 0.04$. The dependence of this $\langle A \rangle_{\text{non-pert.}}$ on W is shown in Fig.3.62, together with the averages $\langle S' \rangle$ and $\langle 1-T \rangle$ /Kramer, 1980/. We see that $\langle A \rangle_{\text{non-pert.}}$ decreases slowly with increasing W. Compared with the perturbative $\langle A \rangle$ calculated from the perturbative distribution in Fig.3.59, which is also shown in Fig.3.62, $\langle A \rangle_{\text{non-pert.}}$ is still larger than $\langle A \rangle$ even at W = 70 GeV. So for energies as low as 40 GeV we expect $d\sigma/dA$ from $q\bar{q}gg$ and $q\bar{q}q\bar{q}$ production to

117

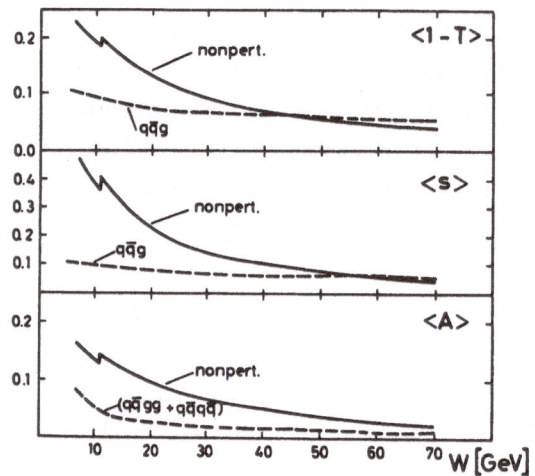

Fig.3.62. Average jet measured <1-T>, <S>, where S = spherocity and <A> for lowest order QCD without fragmentation compared to non-perturbative contributions from $q\bar{q}$ final state including effects of weak decays as a function of W

be larger than the non-perturbative $d\sigma/dA$ from $q\bar{q}$ production only for rather large A. Additional background comes from fragmentation corrections to the 3-jet cross section: $e^+e^- \to q\bar{q}g$. Although this contribution is reduced in magnitude by a factor $\simeq 10$, if compared to the 2-jet contribution, it is broader than the 2-jet contribution. The 3-jet contribution is calculated with a thrust cut-off $T_0 = 0.9$ motivated by demanding $\sigma(q\bar{q}g)/\sigma^{(2)} \simeq \alpha_s$. Instead to compare with the tree-graph prediction of Fig.3.59 the 4-jet contribution is supplemented with quark and gluon fragmentation based on the independent fragmentation model of Ali et al. (see Sect.2.4). The result for an acoplanarity cut-off $A_0 = 0.05$ is shown in Fig.3.61 also. We notice the big change of the perturbative distribution caused by the fragmentation of quarks and gluons. At small $A < A_0$ the distribution is reduced by the cut-off. In this region it is replaced by the $q\bar{q}$ and the $q\bar{q}g$ contribution. At larger A the distribution is enhanced. The normalization of the three curves in Fig.3.61 is such, that the area under the sum of the three contributions is one. The relative normalization of the 2-jet, 3-jet and 4-jet cross sections is 0.83, 0.13 and 0.05. Figure 3.61 shows that A must be larger than 0.3 before the 4-jet contribution exceeds the background from $q\bar{q}g$. Therefore this would be the region, where the perturbative 4-jet contribution can be tested, i.e. where genuine 4-jet events originating from $q\bar{q}gg$ final states can be discovered. In this region $(1/\sigma^{(2)})d\sigma/dA$ decreased by two orders of magnitude. Thus high-statistics experiments are needed to see the α_s^2 term. However, applying a restrictive thrust cut would eliminate almost the total 2-jet signal without losing any 4-jet events. For example, considering only events with $T \leq 0.75$ leads to

$$\sigma^{2\text{-jet}} : \sigma^{3\text{-jet}} : \sigma^{4\text{-jet}} = 0.02 : 0.66 : 0.32$$

so the 4-jet term is very much enhanced.

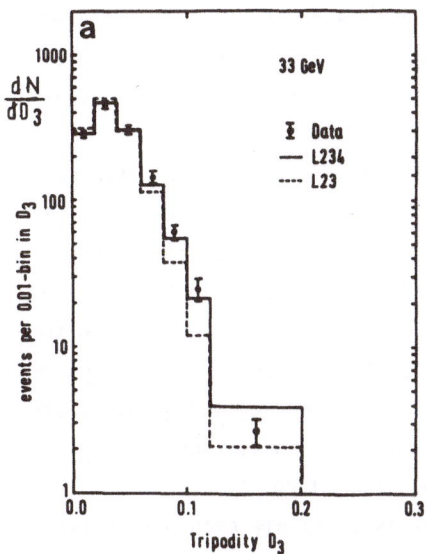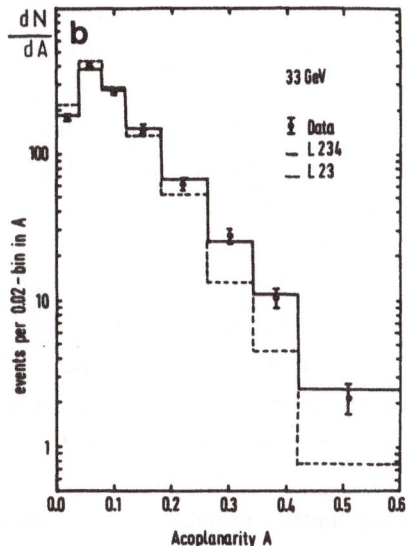

Fig.3.63. Experimental distributions a) dN/dD_3 and b) dN/dA for all events from the JADE Collaboration. D_3 and A are calculated from measured hadron momenta. Histograms represent QCD predictions without (L_{23}) and with (L_{234}) $q\bar{q}gg + q\bar{q}q\bar{q}$ final states ($\lambda = 8\%$) including fragmentation

Results of an analysis aimed at showing evidence for the production of genuine 4-jet events has been published recently by the JADE Collaboration /Bartel et al., 1982b/. The A and D_3 distributions which they obtained for $W = 33$ GeV are shown in Fig.3.63. D_3 stands for tripodity which was defined in Sect.2.3 and which was invented by Nachtmann and Reiter /1982a/ to isolate contributions with the 3-gluon coupling ($D_3 > 0$). The experimental data which are based on the momenta of measured hadrons (not partons) are compared with two models. The model L_{234} consists of $q\bar{q}$, $q\bar{q}g$, $q\bar{q}gg$ and $q\bar{q}q\bar{q}$ final states which all fragment into hadrons according to the Lund scheme. L_{23} contains only $q\bar{q}$ and $q\bar{q}g$ contributions. We see that the model L_{234} describes the data much better than L_{23}. For $A > 0.25$ and $D_3 > 0.1$ the data points lie more than a factor of 2 higher than the predictions of the model L_{23} in agreement with the expectations based on the results in Fig.3.61.

The results presented in Fig.3.63 are not a speciality of the Lund fragmentation model which is based on strings. (For the implementation of the string picture in 4-jet production see Gustafson /1982/.) Comparisons with an independent fragmentation model, like the Ali model, lead to the same conclusions. Furthermore the authors investigated in detail, whether the large tail of the D_3 and A distributions can be explained by varying the fragmentation parameters in the L_{23} model, in particular in the fragmentation of heavy b quarks. This could be ruled out, if one demanded that the distributions at small D_3 and A remain unchanged. The same analysis was performed at W = 22 GeV. It was found that at this energy the data could be well explained by the L_{23} model. This means that the background coming from $q\bar{q}$ and $q\bar{q}g$, because of the non-perturbative jet broadening, is still too large at larger D_3 and A, to see the contribution of 4 parton final states hidden in the tail.

The amount of genuine 4-jet production was also determined. Introducing the contribution of $q\bar{q}gg$ and $q\bar{q}q\bar{q}$ with a free normalization parameter λ into the model, it was found that this λ was definitely unequal zero. In particular, the data in Fig. 3.63 yield $\lambda = (8.2 \pm 1.2)\%$ with an A cut-off of $A_0 = 0.05$ introduced in the 4-parton cross sections. This has to be compared with $\lambda = 5\%$ mentioned in connection with the theoretical curves in Fig.3.61 which were calculated with $\alpha_s = 0.2$.

So we conclude that genuine 4-jet contributions of the form $q\bar{q}gg$ and $q\bar{q}q\bar{q}$, as predicted by QCD, have been discovered at W = 33 GeV with a rate consistent with a coupling constant $\alpha_s = 0.2$ known from lowest order 3-jet production.

For studying further properties of 4-jet final states certainly even higher statistics and/or higher energies are needed. Proposals for such measurements can be found in the literature, for example, in the work of Ali et al. /1980/; Körner, Schierholz and Willrodt /1981/; Chandramohan and Clavelli /1980,1981/; Nachtmann and Reiter /1982b/ and Clavelli and von Gehlen /1982/.

We may ask, for what reason are we allowed to stop with 4 jets and not to establish the contributions of 5, 6, ... jets which certainly exist as well. From our considerations it should be clear, that final states with more than 4 jets have an even larger background produced by fragmentation effects of $q\bar{q}$, $q\bar{q}g$, $q\bar{q}gg$ and $q\bar{q}q\bar{q}$ final states. To suppress this background one needs even higher energies or very large statistics in order to detect these multi jets.

A second point concerns the problem of calculating the genuine 4-jet rate from the tree graphs of Fig.3.37. For a complete calculation it is necessary to add the two-loop contributions and the infrared singular contributions of 5-parton states $q\bar{q}ggg$, $q\bar{q}q\bar{q}g$ of order α_s^3. This is connected again with parameters y or ε, δ, respectively, to separate 4 and 5 jet final states. These parameters should be chosen of the same magnitude, as we did it for the separation of 3 and 4 jets or of 2 and 3 jets, respectively. In this case we expect that the 4-jet rate differs only very little from the rate calculated in Born approximation, i.e. in order α_s^2. This means, only with such a choice of resolution parameters we can interpret results for multi-jets obtained in tree approximation without the necessity of going to higher and higher orders of perturbation theory.

3.2.4   Renormalization Scheme Dependence

In Sect.3.2.2 we emphasized that the higher order correction for the 3-jet cross section is calculated for the $\overline{MS}$ renormalization scheme. In this section we shall explain in more detail how these different renormalization schemes are defined and how they are related. Furthermore we want to understand, why some schemes are more sensible than others.

To perform the renormalization properly, it is necessary to introduce a regularization procedure into QCD. For perturbation theory the dimensional regularization of t'Hooft and Veltman is the most convenient one, since it respects gauge invariance and makes it unnecessary to introduce extra counter terms for recovering gauge invariance. In n ≠ 4 dimensions the QCD coupling constant g has a dimension which is factored out as μ^{ε}, so that the remaining coupling becomes dimensionless: $g \rightarrow \mu^{\varepsilon}g$ (2ε = 4-n). This way an arbitrary dimensional parameter, the scale μ, is introduced into the theory and g becomes dependent on μ and we write $g = g(\mu^2)$ and $\alpha_s = \alpha_s(\mu^2)$. In lowest order of α_s this scale remains undefined. This means, in this order it is not known for which μ the α_s determined from experimental data is obtained. Only in higher orders $O(\alpha_s^2)$ the scale μ is fixed in the way as we have explained it with the α_s^2 corrections to the 3-jet cross section in Sect.3.2.2.

If one goes to higher orders of perturbation theory the quark-gluon coupling g must be renormalized. Let g_b be the bare or unrenormalized coupling and g_r the renormalized coupling then both are related by

$$g_r = Z_3^{1/2}(\varepsilon,\mu)Z_2(\varepsilon,\mu)Z_1^{-1}(\varepsilon,\mu)\mu^{-\varepsilon}g_b \quad . \tag{3.2.32}$$

In (3.2.32) Z_1 is the renormalization constant for the quark-antiquark-gluon vertex, Z_2 for the quark field and Z_3 for the gluon field. These renormalization constants are calculated from corresponding Feynman diagrams. Assuming that quarks and gluons are on their mass shell ($p^2 = 0$) the result of the calculation depends whether n > 4 or n < 4. For the renormalization only the pole terms in ε which occur for n < 4, i.e. which come from the ultraviolet divergence of the integrals, are of interest. These are the pole terms in ε_{UV}. The other contribution is obtained for n > 4 and the complete result can be written as /Jones, 1974; Caswell, 1974; Fabricius, Kramer, Schierholz and Schmitt, 1981/:

$$Z_1 = 1 - \frac{g_r^2}{16\pi^2} (N_c + C_F)\left(\frac{1}{\varepsilon_{UV}} - \frac{1}{\varepsilon_{IR}}\right)$$

$$Z_2 = 1 - \frac{g_r^2}{16\pi^2} C_F\left(\frac{1}{\varepsilon_{UV}} - \frac{1}{\varepsilon_{IR}}\right) \tag{3.2.33}$$

$$Z_3 = 1 + \frac{g_r^2}{16\pi^2} \left(\frac{5}{3} N_c - \frac{2}{3} N_f \right) \left(\frac{1}{\varepsilon_{UV}} - \frac{1}{\varepsilon_{IR}} \right) \quad .$$

In the minimal subtraction scheme only the pole terms proportional to ε_{UV}^{-1} are absorbed in the coupling constant so that the following relation between g_r and g_b follows from (3.2.32):

$$g_r(MS,\mu) = \left[1 + \frac{g_r^2}{16\pi^2} \left(\frac{11}{3} N_c - \frac{2}{3} N_f \right) \frac{1}{2\varepsilon_{UV}} \right] \mu^{-\varepsilon} g_b \quad . \tag{3.2.34}$$

The second term in (3.2.34) determines the subtraction term in higher order $e^+e^- \to q\bar{q}g$ which is proportional to the $O(\alpha_s)$ matrix element $B^V + \varepsilon B^S$ [see (3.1.62)]. From (3.2.34) and the condition that g_b is independent of μ the well-known renormalization group equation /Stueckelberg and Petermann, 1953; Gell-Mann and Low, 1954/ for $g_r(\mu)$ is derived

$$\mu \frac{dg_r(\mu)}{d\mu} = - \beta_0 g_r^3(\mu) \tag{3.2.35}$$

where

$$\beta_0 = \frac{1}{16\pi^2} \left(\frac{11}{3} N_c - \frac{2}{3} N_f \right) \quad . \tag{3.2.36}$$

Integrating this equation we obtain the scale dependence of $\alpha_s = g_r^2/4\pi$ in the form (3.2.8) or with the boundary condition $g_r(\mu=\Lambda) = \infty$ in the well-known form

$$g_r^2(q^2) = \frac{1}{\beta_0 \ln(q^2/\Lambda^2)} \tag{3.2.37}$$

which defines the scale parameter Λ in the one-loop approximation. Taking also two-loop contributions into account (3.2.37) is replaced by

$$g_r^2(q^2) = \frac{1}{\beta_0 \ln(q^2/\Lambda^2) + (\beta_1/\beta_0) \ln\ln(q^2/\Lambda^2)} \tag{3.2.38}$$

where

$$\beta_1 = \frac{1}{(16\pi^2)^2} \left(\frac{34}{3} N_c^2 - \frac{10}{3} N_c N_f - 2 C_F N_f \right) = \frac{1}{256\pi^4} \left(102 - \frac{38}{3} N_f \right) \quad . \tag{3.2.39}$$

In case $g_r^2(q^2)$ is known for a given q^2 we can instead of g_r^2 state the value of Λ as we have done it at the end of Sect.3.2.2 which enables us to compare coupling constants obtained at different q^2. The relation (3.2.34) defines the renormalized

coupling for the minimal subtraction scheme (MS). The Λ calculated from such a defined coupling g_r(MS) is the corresponding Λ_{MS}.

Obviously this definition of the renormalization of g is not unique. It is pos-
sible to absorb arbitrary finite terms (for $\varepsilon \to 0$) in (3.2.34) in the definition of
g_r. For example, it was discovered that it is advantageous to absorb the constant
$\ln(4\pi)-\gamma$. This means, in (3.2.34), we must replace

$$\frac{1}{\varepsilon_{UV}} \to \frac{1}{\varepsilon_{UV}} + [\ln(4\pi)-\gamma] \quad . \tag{3.2.40}$$

The constant $\ln(4\pi)-\gamma$ is an artifact of the dimensional regularization without any
physical significance. Therefore it makes sense to absorb it in the coupling /Bar-
deen, Buras, Duke and Muta, 1978/. The relation (3.2.34) with (3.2.40) defines the
$\overline{MS}$ renormalization scheme which we used in Sect.3.2.2.

A completely different renormalization method is the so-called momentum subtraction scheme, first introduced by Celmaster and Gonsalves /1979a/. In this scheme the
vertices and the coupling are fixed for a space-like momentum point $p^2 = -\mu^2$. This
determines the renormalization constants $Z_1(\varepsilon,\mu)$, $Z_2(\varepsilon,\mu)$ and $Z_3(\varepsilon,\mu)$ for momentum
subtraction. Unfortunately this is not unique since there are several possibilities
for choosing vertices: (i) ggg- vertex $[Z_{1g}(\varepsilon,\mu)]$, (ii) q$\bar{q}$g-vertex $[Z_{1F}(\varepsilon,\mu)]$ or
(iii) GGg-vertex $[Z_{1G}(\varepsilon,\mu)]$. Here GGg is the ghost-ghost-gluon vertex appearing in
covariant gauges (see Sect.1.2). These three possibilities define the so-called MOM,
MOM' and MOM" renormalization. In addition the renormalization depends on the gauge,
for which the Z_i are calculated. This gauge dependence is absent in the $\overline{MS}$ scheme.

For each of these possible renormalizations the coupling $\alpha_s(q^2)$ can be determined
by fitting cross sections calculated at least up to $O(\alpha_s^2)$ with these renormalizations
to experimental data. From this the corresponding Λ is obtained via (3.2.38). Since
the α_s will differ for these various schemes, also different Λ's will be deduced which
are distinguished by Λ_{MS}, $\Lambda_{\overline{MS}}$, Λ_{MOM}, $\Lambda_{MOM'}$ and $\Lambda_{MOM"}$. Relations between these different Λ's have been deduced by Celmaster and Gonsalves /1979/. These relations are val-
id to all orders of perturbation theory under the condition that in the integration
of the renormalization group equation (3.2.35) (extended to higher orders) identical
assumptions about the integration constants are made. As an example we have given the
conversion factors between the Λ's for the renormalizations MS, $\overline{MS}$, MOM, MOM' and MOM"
in Table 3.4 for N_f = 3 and 5 for two gauges, the Landau and Feynman gauge. We see
that the various Λ_{MOM}'s are approximately a factor of two larger than $\Lambda_{\overline{MS}}$. The exact
value depends on the gauge, the N_f value and the vertex chosen for defining the mo-
mentum subtraction. With the help of this table the $\Lambda_{\overline{MS}}$ = 300 MeV obtained in Sect.
3.2.2 can be converted into the MOM scheme. One obtains Λ_{MOM} = 555 MeV for Landau
gauge and Λ_{MOM} = 453 MeV for the Feynman gauge and N_f = 5.

Table 3.4. Conversion factors for Λ's between MS, $\overline{\text{MS}}$, MOM, MOM' and MOM" renormalization scheme for Landau and Feynman gauge and N_f = 3 and 5

Gauge	N_f	$\overline{\text{MS}}$/MS	MOM/$\overline{\text{MS}}$	MOM'/$\overline{\text{MS}}$	MOM"/$\overline{\text{MS}}$
Landau	3	2.66	2.46	2.10	2.33
	5	2.66	1.85	2.07	2.33
Feynman	3	2.66	2.07	1.83	2.69
	5	2.66	1.51	1.76	2.76

As our last point of this section we shall study the influence of changes of the renormalization of α_s on physically measurable quantities, like cross sections etc. Of course, such physical quantities should be independent of the renormalization convention used. But this can be expected only if the perturbation series were known to all orders. In practice, however, one can calculate the first two or three orders of the expansion. The truncated series therefore differ from each other by terms of the first uncomputed order in g, so that the result becomes scheme dependent. For selected physical quantities this dependence is somewhat less pronounced in some special schemes than in others. In the literature one can find criteria which allow to find the "optimal" renormalization scheme for a specific physical quantity /Stevenson, 1981,1982; Pennington, 1982; Kubo and Sakakibara, 1982; Duke and Kimel, 1982/. This is a wide field and has not been applied yet to jet physics. Therefore we shall not consider it here. More as a pedagogical example we shall study, how the renormalization conventions MS, $\overline{\text{MS}}$, MOM and MOM' effect the perturbative prediction for σ_{tot} or R, respectively.

The leading perturbative result for R is (3.1.71). The next-to-leading order terms have been calculated by Dine and Sapirstein /1979/, Chetyrkin, Kataev and Tkachov /1979/ and Celmaster and Gonsalves /1980/ so that

$$R = \left(3 \sum_f Q_f^2\right) \left[1 + \frac{\alpha_s(q^2)}{\pi} + K(RS)\left(\frac{\alpha_s(q)}{\pi}\right)^2 + \dots\right] \tag{3.2.41}$$

The coefficient K(RS) depends on the renormalization scheme (RS). Its derivation will be considered in the next section. The results for the various schemes are:

$$
\begin{aligned}
K(\text{MS}) &= 7.36 - 0.44 \ N_f \\
K(\overline{\text{MS}}) &= 1.985 - 0.115 \ N_f \\
K(\text{MOM}) &= -4.64 + 0.74 \ N_f \\
K(\text{MOM}') &= -2.19 + 0.16 \ N_f \quad .
\end{aligned}
\tag{3.2.42}
$$

We see that only in the MS scheme the higher order correction is large, whereas in the three other schemes the series appears satisfactorily "convergent". Let us assume that the (fictitious) measured value of R = 3.86 at W^3 = 1000 GeV^2 (according to Sect.2.1 the experimental value is R = 3.93 ± 0.10 at W = 34 GeV) is used to determine α_s(MS) from (3.2.41) together with K(MS) in (3.2.42). The result is α_s(MS) = 0.136 which yields Λ_{MS} = 0.166 GeV according to (3.2.38). From this we find with the help of Table 3.4, assuming N_f = 5 and Landau gauge the Λ values for the other schemes $\overline{MS}$, MOM and MOM'. From these we obtain coupling constants α_s according to (3.2.38) and with them the R values in these schemes from (3.2.41) and (3.2.42). These results can be found in Table 3.5. They show that R differs very little from these four schemes. The difference in R is less than 0.5% although the couplings α_s in these schemes change by almost 50% and the Λ's by more than a factor of 5. The change of α_s in the various conventions is to a large extent compensated by the change of the coefficient K in (3.2.41). Thus R is independent of the schemes considered in this exercise.

That R changes very little when we adopted another scheme certainly is caused by the fact that the coefficient K(RS) and also α_s(RS)/π for all schemes RS are very small so that the higher order corrections K(RS) α_s(RS)/π produced only a small shift of R for all schemes RS (compare third and fourth column in Table 3.5). From this we conclude quite generally that only such renormalization conventions should be adopted, which produce small contributions for the higher order corrections of

Table 3.5. Results for R in the renormalization schemes MS, $\overline{MS}$, MOM and MOM' and corresponding Λ and α_s values

Ren. Scheme	Λ GeV	α_s/π	$\frac{\alpha_s}{\pi}(1+K\frac{\alpha_s}{\pi})$	R
MS	0.166	0.0434	0.0531	3.86
$\overline{MS}$	0.442	0.0524	0.0563	3.87
MOM	0.817	0.0605	0.0570	3.88
MOM'	0.914	0.0623	0.0569	3.88

all, or at least, as many as possible, physical quantities. Of course, this "fastest apparent convergence criterion" is not very quantitative, but seems to work as well as more sophisticated convergence criteria /Pennington, 1982/.

The scheme in which the higher order corrections vanish is the so-called low-order scheme: Let us consider a physical quantity, like R, and let us shift α_s by a change of the renormalization scheme into α_s', where α_s and α_s' are related by

$$\alpha_c = \frac{\alpha_s}{\pi} = \alpha_c' (1 + \kappa \alpha_c' + \dots) \quad . \tag{3.2.43}$$

This causes a shift of R since K in (3.2.41) is changed into

$$K' = K + \kappa . \tag{3.2.44}$$

If we choose $\kappa = -K$, then in the new scheme $K' = 0$, i.e. R is fully determined by the lowest order term $R = (3 \sum_f Q_f^2)(1+\alpha_c')$. For our example in Table 3.5 the corresponding α_s is $\alpha_s(\text{low}) = 0.166$. This does not mean that other physical quantities are also given by the lowest order perturbation theory. The contrary may happen, other physical quantities may have large higher order corrections in this particular low-order scheme. In this case the low-order scheme deduced from the R value is not useful.

In connection with the abelian gluon vector theory discussed in Sect.3.2.2 we had introduced such a low-order scheme for the 3-jet cross section $(1/\sigma)d\sigma/dT$. With this we could fit the 3-jet cross section and find the corresponding α_A from measured data. But the perturbation theory of R in this scheme was very bad [see (3.2.20)].

3.2.5 Total Cross Section up to $O(\alpha_s^2)$

It would be very instructive to calculate the total annihilation cross section σ_{tot} by summing up the 2-, 3- and 4-jet cross sections. This is not possible yet, since the 2-jet cross section has not been calculated up to the order α_s^2. However, σ_{tot} can be computed more easily by summing over all final states from the beginning. σ_{tot} is related to the imaginary part of the vacuum polarization or the inverse photon propagator. Thus by computing the inverse photon propagator up to $O(\alpha_s^2)$ one can deduce $\sigma_{tot}(q^2)$. This was done the first time by Dine and Sapirstein /1979/, Chetyrkin, Kataev and Tkadov /1979/ and by Celmaster and Gonsalves /1979/.

In the following we shall outline the calculation of the inverse photon propagator $\Gamma_{\mu\nu}(q)$ up to $O(\alpha_s^2)$ but in lowest order of α. $\Gamma_{\mu\nu}(q)$ depends only on the momentum of the virtual photon q and has, because of gauge invariance, the following decomposition

$$\Gamma_{\mu\nu}(q) = (-g_{\mu\nu} + q_\mu q_\nu/q^2)\Pi(q^2) \quad . \tag{3.2.45}$$

First $\Gamma_{\mu\nu}(q)$ is calculated for space-like $q^2 < 0$ which by analytic continuation yields $\Pi(q^2)$ for time-like $q^2 > 0$. σ_{tot} is obtained from the imaginary part of $\Pi(q^2)$ by

$$\sigma_{tot}(q^2) = \frac{4\pi\alpha}{q^2} \, \text{Im} \, \Pi(q^2) \quad . \tag{3.2.46}$$

The computation of $\Pi(q^2)$ can be simplified with the help of the renormalization group. $\Pi(q^2)$ has the naive dimension 2. Therefore we have for the scale dependence of $\Pi(q^2)$ the following renormalization group equation /Callan, 1970; Symanzik, 1970/:

$$\left(-\frac{\partial}{\partial t} + \beta(e)\frac{\partial}{\partial e} + \beta(g)\frac{\partial}{\partial g} + 2 + \gamma_\gamma\right)\Pi(q^2) = 0 \tag{3.2.47}$$

where $t = (1/2)\ln(-q^2/\mu^2)$. $\beta(e)$ and $\beta(g)$ are the β functions which determine the scale dependence of the electromagnetic coupling and the quark-gluon coupling [see (3.2.35) for g]. γ_γ is the anomalous dimension of the photon. The solution of (3.2.47) is

$$\Pi(q^2) = (-q^2)\exp\left[\int_0^t dt' \gamma_\gamma(\bar{e}(t'),\bar{g}(t'))\right] \quad . \tag{3.2.48}$$

The t dependence of $\bar{g}(t)$ is known through (3.2.35) which has the solution (3.2.8), whereas the t dependence of $\bar{e}(t)$ will be neglected, since we restrict ourselves to the lowest order in α. Therefore $\Pi(q^2)$ is fully known, if the anomalous dimension γ_γ has been calculated. γ_γ is related to the renormalization constant Z_γ of the photon propagator by

$$\gamma_\gamma = -\lim_{\varepsilon \to 0} \mu\frac{\partial\ln Z_\gamma}{\partial\mu} \quad . \tag{3.2.49}$$

Z_γ is calculated with n dimensional regularization so that $4-n = 2\varepsilon$ as before.

Let us demonstrate the calculation with the simplest example, the one-loop graph in Fig.3.64. For this diagram Z_γ is

$$Z_\gamma = 1 - \frac{e^2}{12\pi^2}\left(3\sum_f Q_f^2\right)\frac{1}{\varepsilon} \quad . \tag{3.2.50}$$

Q_f is the charge of the quark with flavour f and e is the renormalized electromagnetic coupling constant $e(\mu)$, which depends on μ. The unrenormalized coupling $e_b = Z_\gamma^{-1}\mu^\varepsilon e$ is independent of μ, from which with (3.2.49) and (3.2.50) we obtain for γ_γ

$$\gamma_\gamma = -\frac{e^2}{6\pi^2}\sum_f 3Q_f^2 \quad . \tag{3.2.51}$$

This is substituted into (3.2.48) with the result

$$\Pi(q^2) = (-q^2)(1 - \frac{2\alpha}{3\pi}t\sum_f 3Q_f^2) \quad . \tag{3.2.52}$$

In (3.2.52) $t = (1/2)\ln(-q^2/\mu^2)$ is continued to time-like $q^2 > 0$ and the imaginary

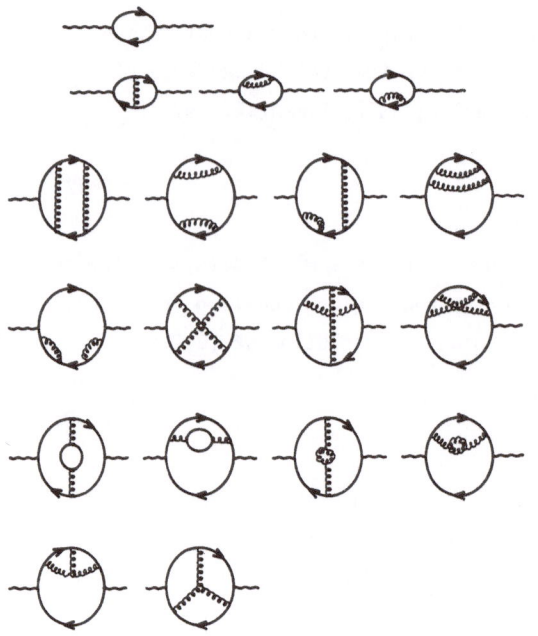

Fig.3.64. Diagrams for vacuum polarization of the photon up to fourth order in g. The eleventh and twelfth diagram of the g^4 diagrams contains also the contributions of the ghosts and the four-gluon coupling

part is taken, from which with (3.2.46) the well-known formula for $\sigma_{tot}(q^2)$ follows

$$\sigma_{tot}(q^2) = \frac{4\pi\alpha^2}{3q^2} \sum_f 3Q_f^2 \quad . \tag{3.2.53}$$

We see that for calculating γ_γ and σ_{tot} we need only the simple poles in ϵ of the loop graphs in Fig.3.64. This constitutes a tremendous simplification.

In the next order the contributions to γ_γ are of order $g^2(t)$ which are calculated from the two-loop diagrams in Fig.3.64. The result is

$$\gamma_\gamma = -\frac{e^2}{6\pi^2} \sum_f 3Q_f^2 \left[1 + \frac{3C_F}{16\pi^2} g^2(t)\right] \quad . \tag{3.2.54}$$

This is substituted into (3.2.48). Then the integration over t' yields

$$\Pi(q^2) = (-q^2) \left\{ 1 - \frac{2\alpha}{3\pi} (\sum_f 3Q_f^2) \left[t + \frac{3C_F}{2\beta_0} \ln\left(1 + \frac{\beta_0}{8\pi^2} g^2(t)\right)\right]\right\} \quad . \tag{3.2.55}$$

In (3.2.55) we have included also the zeroth-order term (3.2.52). β_0 is defined differently than in (3.2.36). Here it is $\beta_0 = 11N_c/3 - 2N_f/3$. Through analytic continuation to positive q^2 the well-known formula is obtained

$$\sigma_{tot}(q^2) = \frac{4\pi\alpha^2}{3q^2} \left(\sum_f 3Q_f^2 \right) \left(1 + \frac{3}{4} C_F \frac{\alpha_s(q^2)}{\pi} \right) \tag{3.2.56}$$

in agreement with (3.1.70) which was obtained from the sum of the 2- and 3-jet cross section. Since (3.2.56) is based on the renormalization group, it follows that $\alpha_s(q^2)$ should be taken at the scale q^2, the center-of-mass energy squared. For our earlier result (3.1.70), which was based on $O(\alpha_s)$ perturbation theory, the scale μ was completely arbitrary. So the renormalization group treatment adjusts the scale to the only momentum scale present, namely q^2.

As we know, up to $O(g^2)$ there is no renormalization of g. Only in order α_s^2 renormalization of g comes in and so σ_{tot}, calculated in this order, must depend on the renormalization scheme. This can be seen directly also when we calculate Z_γ in higher order of g. Let us discuss this in connection with the MS and $\overline{MS}$ scheme. Quite generally Z_γ has the following structure, assuming the MS scheme for renormalization first

$$Z_\gamma^{MS} = 1 + e^2 \sum_{l=1}^{\infty} \left(g_{MS}^2 \right)^{l-1} \sum_{k=1}^{l} Z_{1,-k} \varepsilon^{-k} \quad . \tag{3.2.57}$$

Changing now to the $\overline{MS}$ scheme, where in addition to the ε^{-1} term also the constant term $\ln(4\pi)-\gamma$ is absorbed into the renormalization coupling $g_{\overline{MS}}^2$ [see (3.2.40)], $Z_\gamma^{\overline{MS}}$ has the following structure

$$Z_\gamma^{\overline{MS}} = 1 + e^2 \sum_{l=1}^{\infty} \left(g_{\overline{MS}}^2 \right)^{l-1} \exp\{(l-1)\varepsilon[\ln(4\pi)-\gamma]\} \left(\sum_{k=1}^{l} Z_{1,-k} \varepsilon^{-k} \right) \quad . \tag{3.2.58}$$

The exponential factor $e^{(l-1)\varepsilon[\ln(4\pi)-\gamma]}$ comes into play only if $l \geq 2$ and if there are pole terms $1/\varepsilon^2$. For $l = 2$ we have $Z_{2,-2} = 0$ because of the gauge invariance of the photon propagator. Therefore, only for $l = 3$, i.e. in order g^4, a change of the renormalization scheme has an effect. Then the renormalization scheme and also the Λ parameter can be defined.

For $l = 3$ we need the coefficient $Z_{3,-1}$. This is calculated from the three-loop diagrams in Fig.3.64. Of course, this is a much more involved computation than the one- and two-loop diagrams. We take the result from the work of Chetyrkin, Kataev and Tkachov /1980/ or that of Celmaster and Gonsalves /1980/:

$$Z_{3,-1} = - \frac{1}{(16\pi^2)^2} \frac{16}{3} Z_{1,-1} K(MS) \quad . \tag{3.2.59}$$

Here $Z_{1,-1}$ is the coefficient of the lowest order result (3.2.50)

$$Z_{1,-1} = - \frac{1}{12\pi^2} \sum_f 3Q_f^2 \quad . \tag{3.2.60}$$

The factor K is the following in the MS renormalization convention

$$K(MS) = C_F \left\{ -\frac{3}{32} C_F - \frac{3}{4} \beta_0 \zeta_3 - \frac{33}{48} N_f + \frac{132}{32} N_c + \frac{3}{16} \beta_0 [\ln(4\pi) - \gamma] \right\} \qquad (3.2.61)$$

with ζ_3 being the number

$$\zeta_3 = \sum_{n=1}^{\infty} \frac{1}{n^3} = 1.2021 \quad . \qquad (3.2.62)$$

From (3.2.57) with (3.2.59) we obtain

$$\gamma_\gamma^{MS} = -6e^2 g_{MS}^4 Z_{3,-1} \qquad (3.2.63)$$

which is needed to calculate $\Pi(q^2)$ and from this σ_{tot}. In the $\overline{MS}$ scheme we start from (3.2.58) to obtain $\gamma_\gamma^{\overline{MS}}$ with the result

$$\gamma_\gamma^{\overline{MS}} = -6e^2 g_{\overline{MS}}^4 \{ Z_{3,-1} + 2[\ln(4\pi) - \gamma] Z_{3,-2} \} \quad . \qquad (3.2.64)$$

The coefficient $Z_{3,-2}$ is

$$Z_{3,-2} = -\frac{\beta_0}{48\pi^2} Z_{2,-1} \qquad (3.2.65)$$

where

$$Z_{2,-1} = -\frac{3C_F}{32\pi^2} Z_{1,-1} \quad . \qquad (3.2.66)$$

Therefore the term proportional to $[\ln(4\pi) - \gamma]$ in K(MS) cancels and $\gamma_\gamma^{\overline{MS}}$ has the same form as (3.2.63) except that K(MS) is replaced by

$$K(\overline{MS}) = C_F \left(-\frac{3}{32} C_F - \frac{3}{4} \beta_0 \zeta_3 - \frac{33}{48} N_f + \frac{123}{32} N_c \right) \quad . \qquad (3.2.67)$$

Then the final formula for σ_{tot} is

$$\sigma_{tot}(q^2) = \frac{4\pi\alpha^2}{3q^2} \sum_f N_c Q_f^2 \left[1 + \frac{3}{4} C_F \frac{\alpha_s(q^2)}{\pi} + K \left(\frac{\alpha_s(q^2)}{\pi} \right)^2 + \dots \right] \quad . \qquad (3.2.68)$$

In this formula K is given by (3.2.61) for σ_{tot} in the MS scheme and by (3.2.67) for the $\overline{MS}$ scheme.

From (3.2.68) we obtain the result (3.2.42) and the result for the abelian vector theory used in Sect.3.2.3 if we substitute $C_F = 1$, $N_c = 0$ and $N_f \rightarrow 6N_f$ in the coefficients of α_s/π and $(\alpha_s/\pi)^2$ in (3.2.68).

We remark again that the coefficient K is of the order of 1 in the renormalization schemes $\overline{MS}$, MOM and MOM', so that in these conventions the higher order contributions to $\sigma_{tot}(q^2)$ are very small, of the order of 0.5% for $\alpha_s/\pi = 0.05$.

From the average experimental value of R in the energy range $30 \leq W \leq 36.7$ GeV which, as reported in Sect.2.1, is equal to (with weak contributions subtracted)

$$R(W = 34 \text{ GeV}) = 3.93 \pm 0.10 \tag{3.2.69}$$

we can calculate the α_s values for this energy in the three schemes $\overline{MS}$, MOM and MOM'. The result for $N_f = 5$ is:

$$(\alpha_s)_{\overline{MS}} = 0.206 \pm 0.074$$
$$(\alpha_s)_{MOM} = 0.243 \pm 0.104 \tag{3.2.70}$$
$$(\alpha_s)_{MOM'} = 0.254 \pm 0.119 \quad .$$

The value for $(\alpha_s)_{\overline{MS}}$ is consistent with the values we obtained from the 3-jet cross section. Although the error of R is less than 3% the error of α_s deduced from R is still large.

This completes the section on $O(\alpha_s^2)$ corrections to jet cross sections in QCD.

4. Summary and Conclusions

For the summary it seems appropriate to remember the characteristic features of QCD which we have collected in Sect.1.3 and to ask which of the points (i) to (vi) we can consider verified through the perturbative analysis of jet phenomena in e^+e^- annihilation. We saw that the spin 1/2 nature for quarks follows from the angular distribution of the jet axis and $N_c = 3$ results from the measurement of R. (ii), the existence of gluons and their vector character together with (iii) the coupling of gluons to quarks is well verified by the analysis of 3-jet final states. The question whether the 3- and 4-gluon couplings exist has been answered by showing that theories with abelian vector gluons can be excluded. What remains is the question concerning the universality of all couplings in the Lagrangian and the verification that the quark-gluon coupling changes with the scale q^2 in a characteristic fashion. To investigate the latter problem one needs higher energies in order to see how α_s decreases with increasing q^2. This will be difficult since at high energies more and more jets will be produced and also the influence of the Z propagator will be stronger at these energies.

In this review we restricted ourselves to a presentation and interpretation of perturbative results in fixed order up to $O(\alpha_s^2)$. This seems the most natural strategy, since there is a relation between the number of jets possible and the perturbative order. Now all jet cross sections are known up to order α_s^2. These results have been used to determine the strong coupling constant α_s employing various methods of analysis. This way one is able to obtain α_s in a specified renormalization scheme and to compare with results coming from analysis of other processes. The value is $\alpha_s = 0.15$. This corresponds to $\Lambda_{\overline{MS}} = 300$ MeV which compares well with determinations of $\Lambda_{\overline{MS}}$ from deep inelastic lepton nucleon scattering experiments.

A consistent interpretation of jet phenomena in the framework of QCD perturbation theory requires an appropriate definition of jets due to the infrared properties of QCD. This is quite analogous to the perturbative treatment of QED. For this definition of jets one needs resolution parameters which determine the jet multiplicities. They must be chosen in such a way that the jet rates diminish with increasing number of jets with the power of α_s.

A quantitative analysis of experimental data requires still fragmentation models as input. They are more or less phenomenological. It is conceivable that in the future these models can be improved and can be put on a more solid theoretical basis. Going to higher center-of-mass energies will have the effect that more than four jets will be produced. This requires the computation of higher orders in perturbation theory, a difficult task. A way out might be the calculation of these higher order terms in the leading logarithm approximation and to use the results of order α_s^2 only in the appropriate kinematic regions. This marriage of leading log and perturbative results might also allow to replace the phenomenological hadronization models in particular kinematical domains.

References

Abers, E.S., B.W. Lee (1973): Phys. Rep. *9*, 1 /Sect.1.2/
Abramowicz, H. et al. (1983): Z. Phys. *C17*, 283 /Sect.3.2.2/
Adeva, B. et al. (1983a): Phys. Rev. Lett. *51*, 443 /Sect.2.4.1/
Adeva, B. et al. (1983b): Phys. Rev. Lett. *50*, 799 /Sect.3.1.9/
Adeva, B. et al. (1983c): Phys. Rev. Lett. *50*, 2051 /Sects.3.1.9, 3.2.2/
Alexander, G. (1978): Proceedings XIXth International Conference on High Energy
    Physics (Tokyo), 255 /Sect.2.3/
Ali, A. (1979): Z. Phys. *C1*, 25 /Sect.2.4.2/
Ali, A. (1982): Phys. Lett. *110*, 67 /Sect.3.2.2/
Ali, A., F. Barreiro (1982): Phys. Lett. *118B*, 155 /Sects.3.1.5, 3.2.2/
Ali, A., J.G. Körner, G. Kramer, J. Willrodt (1979a): Z. Phys. *C1*, 203 /Sects.2.4.2,
    3.1.9/
Ali, A., J.G. Körner, G. Kramer, J. Willrodt (1979b): Z. Phys. *C1*, 269 /Sects.2.4.2,
    3.1.9/
Ali, A., J.G. Körner, G. Kramer, J. Willrodt (1979c): Z. Phys. *C2*, 33 /Sect.2.4.2/
Ali, A., J.G. Körner, G. Kramer, J. Willrodt (1979d): Phys. Lett. *83B*, 375
    /Sect.2.4.2/
Ali, A., J.G. Körner, G. Kramer, J. Willrodt (1980): Nucl. Phys. *B168*, 409
    /Sects.2.4.2, 3.1.9/
Ali, A. et al. (1979): Phys. Lett. *82B*, 285 /Sects.3.2.1,3/
Ali, A. et al. (1980): Nucl. Phys. *B167*, 454 /Sects.3.2.1,3/
Ali, A., G. Kramer, E. Pietarinen, J. Willrodt (1979): DESY-Report 79/86 /Sect.3.1.9/
Ali, A., G. Kramer, E. Pietarinen, J. Willordt (1980): Phys. Lett. *93B*, 155 /Sects.
    2.4.2, 3.1.9/
Altarelli, G. (1982): Phys. Rep. *81*, 1 /Preface/
Altarelli, G., G. Parisi (1977): Nucl. Phys. *B126*, 298 /Sect.2.4.2/
Althoff, M. et al. (1983a): DESY-Report 83-010 /Sect.2.4.1/
Althoff, M. et al. (1983b): DESY-Report 83-130, Z. Phys. (to be published)
Andersson, B., G. Gustafson, C. Peterson (1979): Z. Phys. *C1*, 105 /Sect.2.4.2/
Andersson, B., G. Gustafson (1980): Z. Phys. *C3*, 223 /Sect.2.4.2/
Andersson, B., G. Gustafson, T. Sjöstrand (1980): Phys. Lett. *94B*, 211; Z. Phys.
    C6, 235 /Sect.2.4.2/
Andersson, B., G. Gustafson, T. Sjöstrand (1982): Z. Phys. *C12*, 49 /Sect.2.4.2/
Andersson, B., G. Gustafson, B. Söderberg (1983): Z. Phys. *C20*, 317 /Sect.2.4.1/
Appelquist, T., H. Georgi (1973): Phys. Rev. *D8*, 4000 /Sect.2.1/
Avram, N.M., D.H. Schiller (1974): Nucl. Phys. *B70*, 272 /Sects.3.1.2,6/

Babcock, J.B., R.E. Cutkosky (1981): Nucl. Phys. *B176*, 113 /Sect.3.1.9/
Babcock, J.B., R.E. Cutkosky (1982a): Z. Phys. *C15*, 133 /Sect.3.1.9/
Babcock, J.B., R.E. Cutkosky (1982b): Nucl. Phys. *B201*, 527 /Sect.3.1.9/
Bäcker, A. (1982): Z. Phys. *C12*, 161 /Sect.3.1.9/
Barber, D.P. et al. (1979): Phys. Rev. Lett. *43*, 830 /Sect.3.1.9/

Bardeen, W.A., A.J. Buras, D.W. Duke, T. Muta (1978): Phys. Rev. *D18*, 3998
/Sect.3.2.4/
Barker, I.S., B.R. Martin, G. Shaw (1983): Z. Phys. *C19*, 147
Bartel, W. et al. (1980): Phys. Lett. *91B*, 142 /Sect.3.1.9/
Bartel, W. et al. (1981): Phys. Lett. *101B*, 129 /Sect.3.1.9/
Bartel, W. et al. (1982a): Phys. Lett. *119B*, 239 /Sects.3.1.9, 3.2.2/
Bartel, W. et al. (1982b): Phys. Lett. *115B*, 338 /Sect.3.2.2/
Bartel, W. et al. (1983): Phys. Lett. *123B*, 460 /Sects.2.4.2, 3.1.9, 3.2.3/
Basham, C.L., L.S. Brown, S.D. Ellis, S.T. Love (1978): Phys. Rev. *D17*, 2298;
Phys. Rev. Lett. *41*, 1585 /Sect.3.1.5/
Basham, C.L., L.S. Brown, S.D. Ellis, S.T. Love (1979): Phys. Rev. *D19*, 2018
/Sect.3.1.5/
Becher, P., M. Böhm, H. Joos (1981): *Eichtheorien der starken und elektroschwachen
Wechselwirkung* (Teubner, Stuttgart) /Sect.1.2/
Behrend, H.J. et al. (1982a): Phys. Lett. *110B*, 329 /Sect.3.1.9/
Behrend, H.J. et al. (1982b): Z. Phys. *C14*, 95 /Sects.3.1.9, 3.2.2/
Behrend, H.J. et al. (1983): Nucl. Phys. *B218*, 269 /Sects.3.1.5,9/
Behrends, S. et al. (1983): Phys. Rev. Lett. *50*, 881 /Sect.2.4.2/
Berger, Ch. et al. (1978): Phys. Lett. *78B*, 176 /Sect.2.2/
Berger, Ch. et al. (1979): Phys. Lett. *86B*, 418 /Sect.3.1.9/
Berger, Ch. et al. (1980a): Phys. Lett. *97B*, 459 /Sect.3.1.9/
Berger, Ch. et al. (1980b): Phys. Lett. *90B*, 312 /Sect.3.1.9/
Berger, Ch. et al. (1981a): Z. Phys. *C8*, 101 /Sect.2.2/
Berger, Ch. et al. (1981b): Phys. Lett. *99B*, 292 /Sect.3.1.9/
Berger, Ch. et al. (1982): Z. Phys. *C12*, 297 /Sect.3.1.7/
Berger, E.L. (1980): Z. Phys. *C4*, 289 /Sect.3.1.9/
Berman, S., J.D. Bjorken, J. Kogut (1971): Phys. Rev. *D4*, 3388 /Sect.1.3/
Bjorken, J.D. (1967): Proceedings of the 1967 International Symposium on Electron
and Photon Interactions at High Energies (SLAC, Stanford, Calif.) 109 /Sect.1.1/
Bjorken, J.D., E.A. Paschos (1969): Phys. Rev. *185*, 1975 /Sect.1.1/
Bjorken, J.D., S. Brodsky (1970): Phys. Rev. *D1*, 1416 /Sect.2.2/
Binetruy, P., G. Giradi (1979): Phys. Lett. *83B*, 382 /Sect.3.1.7/
Bollini, C.G., J.J. Giambiagi (1972): Nuovo Cim. *12B*, 20 /Sect.3.1.1/
Bopp, F.W. (1979): Z. Phys. *C3*, 171 /Sect.2.3/
Bopp, F.W., D.H. Schiller (1980): Z. Phys. *C6*, 161 /Sect.3.1.2/
Bowler, M.G. (1981): Z. Phys. *C11*, 169 /Sect.2.4.1/
Brandelik, R. et al. (1979): Phys. Lett. *86B*, 243 /Sect.3.1.9/
Brandelik, R. et al. (1980a): Phys. Lett. *94B*, 437 /Sects.2.4.2, 3.1.9/
Brandelik, R. et al. (1980b): Phys. Lett. *97B*, 453 /Sect.3.1.9/
Brandelik, R. et al. (1980c): Z. Phys. *C4*, 237 /Sect.3.1.9/
Brandelik, R. et al. (1980d): Phys. Lett. *92B*, 199 /Sect.2.1/
Brandelik, R. et al. (1981): Phys. Lett. *100B*, 357 /Sect.2.2/
Brandelik, R. et al. (1982): Phys. Lett. *114B*, 65 /Sect.3.1.9/
Brandt, S., H.J. Dahmen (1979): Z. Phys. *C1*, 61 /Sect.2.3/
Brandt, S., Ch. Peyron, R. Sosnovski, A. Wroblewski (1964): Phys. Lett. *12*, 57
/Sect.2.3/
Brown, L.S., S.D. Ellis (1981): Phys. Rev. *D24*, 2383 /Sect.3.1.5/
Brown, L.S., S.P. Li (1982): Phys. Rev. *D26*, 570 /Sect.3.1.5/
Buras, A.J. (1980): Rev. Mod. Phys. *52*, 199 /Preface/
Burrows, C.J. (1979): Z. Phys. *C2*, 215 /Sect.2.4.1/

Cabibbo, N., Parisi, M. Testa (1970): Lett. Nuovo Cim. *4*, 35 /Sect.1.3/
Callan, C.G., D.J. Gross (1969): Phys. Rev. Lett. *22*, 156 /Sect.2.2/
Callan, C.G. (1979): Phys. Rev. *D2*, 1541 /Sect.3.2.5/
Caswell, W. (1974): Phys. Rev. Lett. *33*, 224 /Sect.3.2.4/
Celmaster, W., R.J. Gonsalves (1979a): Phys. Rev. *D20*, 1429 /Sect.3.2.4/
Celmaster, W., R.J. Gonsalves (1979b): Phys. Rev. Lett. *44*, 560; Phys. Rev. *D21*,
3112 /Sects.3.2.4,5/

Chandramohan, T., L. Clavelli (1980): Phys. Lett. *94B*, 409 /Sect.3.2.3/
Chandramohan, T., L. Clavelli (1981): Nucl. Phys. *B185*, 365 /Sect.3.2.3/
Chetyrkin, K.G., A.L. Kataev, F.V. Tkachov (1979): Phys. Lett. *85B*, 277
    /Sects.3.2.4,5/
Chetyrkin, K.G., A.L. Kataev, F.V. Tkachov (1980): Preprint Moscow, INR-P-0170
    /Sect.3.2.5/
Clavelli, L., G. v. Gehlen (1983): Phys. Rev. *D27*, 2063 /Sect.3.2.3/
Clegg, A.B., A. Donnachie (1982): Z. Phys. *C13*, 71 /Sect.2.4.1/
Criegee, L., G. Knies (1982): Phys. Rep. *83*, 151 /Sect.3.1.9/

Daum, H.J., J. Bürger, H. Meyer (1981): Z. Phys. *C8*, 167 /Sect.3.1.9/
De Grand, T.A., Y.J. Ng, S.H.H. Tye (1977): Phys. Rev. *D16*, 3251 /Sects.3.1.9,3.2.3/
De Rujula, A., J. Ellis, E.G. Floratos, M.K. Gaillard (1978): Nucl. Phys. *B138*, 387
    /Sects.3.1.3,4,9/
Deveto, A., D.W. Duke, J.W. Owens, R.G. Roberts (1983): Phys. Rev. *D27*, 508
    /Sect.3.2.2/
Dine, M., J. Sapirstein (1979): Phys. Rev. Lett. *43*, 668 /Sects.3.2.4,5/
Dokshitser, Y.L., D.I. D'yakonov, S.I. Troyan (1980): Phys. Rep. *58*, 269 /Preface/
Dorfan, J. (1981): Z. Phys. *C7*, 349 /Sect.3.1.9/
Drell, S., D. Levy, T. Yan (1970): Phys. Rev. *D1*, 1617 /Sect.2.1/
Duinker, P. (1982): Rev. Mod. Phys. *54*, 325 /Sect.3.1.9/
Duke, D.W., J.D. Kimel (1982): Phys. Rev. *D25*, 2960 /Sect.3.2.4/

Eichmann, G., F. Steiner (1979): Z. Phys. *C1*, 363 /Sect.2.3/
Ellis, J., M.K. Gaillard, G.G. Ross (1976,1977): Nucl. Phys. *B111*, 253 (Erratum
    B130, 516) /Sects.3.1.2,8,9/
Ellis, J., I. Karliner (1979): Nucl. Phys. *B148*, 141 /Sect.3.1.8/
Ellis, J., C.T. Sachrajda (1980): Proceedings of the Cargèse Summer Institute on
    Quarks and Leptons (Plenum, New York) /Preface/
Ellis, R.K., D.A. Ross, E.A. Terrano (1980): Phys. Rev. Lett. *45*, 1226 /Sect.3.2.2/
Ellis, R.K., D.A. Ross (1981): Phys. Lett. *106B*, 88 /Sect.3.2.2/
Ellis, R.K., D.A. Ross, E.A. Terrano (1981): Nucl. Phys. *B178*, 421 /Sects.3.2.2,3/
Elsen, E. (1981): Interner Bericht DESY F22-81/02 /Sects.2.2, 3.1.9/
Engels, J., J. Dabkowski, K. Schilling (1980): Z. Phys. *C3*, 371 /Sect.2.3/

Fabricius, K., I. Schmitt, G. Schierholz, G. Kramer (1980): Phys. Lett. *97B*, 431
    /Sect.3.2.2/
Fabricius, K., I. Schmitt, G. Kramer, G. Schierholz (1980): Phys. Rev. Lett. *45*,
    867 /Sect.3.1.6/
Fabricius, K., G. Kramer, G. Schierholz, I. Schmitt (1982): Z. Phys. *C11*, 315
    /Sects.3.2.2,4/
Farhi, E. (1977): Phys. Rev. Lett. *39*, 1587 /Sect.2.3/
Felst, R. (1981): Proceedings of the 1981 International Symposium on Lepton and
    Photon Interactions at High Energies (Bonn) /Sect.2.1/
Feynman, R.P. (1972): *Photon-Hadron-Interactions* (Benjamin, Reading) /Sects.1.1, 2.1/
Field, R.D. (1978): Conference on QCD at La Jolla, AIP Conference Proc. *55*, 97
    /Preface/
Field, R.D., R.P. Feynman (1978): Nucl. Phys. *B136*, 1 /Sect.2.4.1/
Field, R.D., S. Wolfram (1983): Nucl. Phys. *B213*, 65 /Sect.2.4.2/
Fox, G.C., S. Wolfram (1980a): Nucl. Phys. *B168*, 285 /Sect.2.4.2/
Fox, G.C., S. Wolfram (1980b): Z. Phys. *C4*, 237 /Sect.3.1.5/
Fritzsch, H., M. Gell-Mann (1972): Proceedings of the 16th International Conference
    on High Energy Physics (Chicago) *2*, 135 /Sect.1.2/

Gaemers, K.J.F., J.A.M. Vermaseren (1980): Z. Phys. *C7*, 81 /Sects.3.2.1,3/
Gatto, R. (1965): Proceedings of the International Symposium on Electron and Photon
    Interactions at High Energies (Hamburg) 106 /Sect.2.2/

Gell-Mann, M., F.E. Low (1954): Phys. Rev. *95*, 1300 /Sect.3.2.4/
Gell-Mann, M. (1962): Phys. Rev. *125*, 1067 /Sect.1.2/
Gell-Mann, M. (1964): Phys. Lett. *8*, 214 /Sect.1.1/
Gell-Mann, M. (1972): Acta Phys. Austr. Suppl. IX, 733 /Sect.1.1/
Georgi, H., M. Machacek (1977): Phys. Rev. Lett. *39*, 1237 /Sect.3.1.4/
Georgi, H., J. Sheiman (1979): Phys. Rev. *D20*, 111 /Sect.3.1.4/
Goddard, M.C. (1981): Rutherford-Appleton Laboratory Report RL-81-069 /Sect.3.1.9/
Gottschalk, T.D. (1982): Phys. Lett. *109B*, 331 /Sect.3.2.2/
Gottschalk, T.D. (1983): Nucl. Phys. *B214*, 201 /Sect.2.4.1/
Grayson, S.L., M.P. Tuite (1982): Z. Phys. *C13*, 337 /Sect.3.1.9/
Greenberg, O.W. (1964): Phys. Rev. Lett. *13*, 598 /Sect.1.1/
Gross, D.J., F. Wilczek (1973): Phys. Rev. Lett. *30*, 1343 /Sect.1.3/
Gustafson, G. (1982): Z. Phys. *C15*, 155 /Sect.3.2.2/
Gutbrod, F., G. Kramer, G. Schierholz (1983): DESY-Report 83-044; Z. Phys. *C21*,
    235 /Sect.3.2.2/

Han, M.Y., Y. Nambu (1965): Phys. Rev. *B139*, 1006 /Sect.1.1/
Hanson, G. et al. (1975): Phys. Rev. Lett. *35*, 1609 /Sects.2.2, 2.4.1/
Hanson, G. et al. (1982): Phys. Rev. *D26*, 991 /Sects.2.2, 2.4.1/
Hirshfeld, A.C., G. Kramer (1974): Nucl. Phys. *B74*, 211 /Sect.3.1.2/
Hirshfeld, A.C., G. Kramer, D.H. Schiller (1974): DESY-Report 74/33 /Sects.3.1.2,6/
Hoyer, P., P. Osland, H.G. Sander, T.F. Walsh, P. Zerwas (1979): Nucl. Phys. *B161*,
    349 /Sects.2.4.2, 3.1.4,8,9/
Hoyer, P. (1980): Acta Phys. Pol. *B11*, 133 /Preface/
Hughes, R.J. (1980): Phys. Lett. *97B*, 246; Nucl. Phys. *B186*, 376 /Sect.1.3/

Itzykson, C., J.B. Zuber (1980): *Quantum Field Theory* (McGraw Hill, New York)
    /Sect.1.2/

Jersak, J., E. Laermann, P.M. Zerwas (1982): Phys. Rev. *D25*, 1218 /Sect.3.1.3/
Johnson, P.W., Wu-ki Tung (1982): Z. Phys. *C13*, 87 /Sect.3.1.3/
Jones, D.R. (1974): Nucl. Phys. *B75*, 531 /Sect.3.2.4/
Jones, L.M., R. Migneron (1983): Phys. Rev. *D27*, 2063 /Sect.2.4.1/

Kinoshita, T. (1960): J. Math. Phys. *3*, 650 /Sect.3.1.7/
Körner, J.G., G. Kramer, G. Schierholz, I. Schmitt, K. Fabricius (1980): Phys. Lett.
    94B, 207 /Sect.3.1.6/
Körner, J.G., G. Schierholz, J. Willrodt (1981): Nucl. Phys. *B185*, 365
    /Sects.3.2.1,3/
Koller, K., H.G. Sander, T.F. Walsh, P.M. Zerwas (1980): Z. Phys. *C6*, 131 /Sect.3.1.3/
Koller, K., D.H. Schiller, D. Wähner (1982): Z. Phys. *C12*, 273 /Sect.3.1.3/
Kramer, G., T.F. Walsh (1973): Z. Phys. *263*, 361 /Sect.2.2/
Kramer, G., G. Schierholz, J. Willrodt (1978,1979): Phys. Lett. *78B*, 249 (Erratum
    80B, 433) /Sects.3.1.2,3/
Kramer, G., G. Schierholz (1979): Phys. Lett. *82B*, 108 /Sect.3.1.9/
Kramer, G. (1980): *Field Theoretical Methods in Particle Physics* (Plenum, New York)
    425 /Preface, Sects.2.4.2, 3.1.8,9, 3.2.3/
Kramer, G., G. Schierholz, J. Willrodt (1980): Z. Phys. *C4*, 149 /Sect.3.1.2/
Kramer, G. (1982): DESY-Report 82-079, published in Proceedings of the XIII. Spring
    Symposium on High Energy Physics (Karl-Marx Universität, Leipzig) /Sect.3.2.2/
Kramer, G. (1983): Interner Bericht DESY T-83-01 /Sect.3.2.2/
Kubo, J., S. Sakakibara (1982): Phys. Rev. *D26*, 3656 /Sect.3.2.4/
Kunzst, Z. (1980): Phys. Lett. *99B*, 429 /Sect.3.2.2/
Kunzst, Z. (1981): Phys. Lett. *107B*, 123 /Sect.3.2.2/

Laermann, E., K.H. Streng, P.M. Zerwas (1980): Z. Phys. *C3*, 289 /Sect.3.1.3/
Laermann, E., P.M. Zerwas (1980): Phys. Lett. *89B*, 225 /Sects.3.1.2,8/
Lampe, B., G. Kramer (1983): Physica Scripta *28*, 585

Lanius, K. (1980): DESY-Report 80/36 /Sect.3.2.9/
Lanius, K., H.E. Roloff, D.H. Schiller (1981): Z. Phys. *C8*, 251 /Sect.3.1.9/
Lee, B.W. (1976): *Methods in Field Theory* (North-Holland, Amsterdam) 79 /Sect.1.2/
Lee, T.D., M. Nauenberg (1966): Phys. Rev. *133*, 1594 /Sect.3.1.7/
Leibrandt, G.A. (1975): Rev. Mod. Phys. *47*, 849 /Sect.3.1.1/

MAC Collaboration (1982), (presented by D. Ritson): Proceedings of the 21st International Conference on High Energy Physics, Journal de Physique *43*, C3-52 /Sect.3.1.9/
Marciano, W.J. (1975): Phys. Rev. *D12*, 3861 /Sect.3.1.1/
Marciano, W.J., H. Pagels (1978): Phys. Rep. *36*, 137 /Preface/
MARK J Collaboration (1982), (presented by J. Burger): Proceedings of the 21st International Conference on High Energy Physics, Journal de Physique *43*, C3-63 /Sect.3.1.9/
Marquardt, W., F. Steiner (1980): Phys. Lett. *93B*, 480 /Sect.3.1.5/
Marshall, R. (1981): EPS Conference on HEP (Lisbon) and Rutherford Appleton Laboratory Report RL-81-087 /Sect.3.1.9/
Mazzanti, P., R. Odorico (1980): Z. Phys. *C7*, 61 /Sect.2.4.2/
Meyer, T. (1982): Z. Phys. *C12*, 77 /Sect.2.4.2/
Montvay, I. (1979): Phys. Lett. *84B*, 331 /Sect.2.4.2/
Mursula, K. (1980): DESY-Report 80/29 /Sect.3.1.9/

Nachtmann, O., A. Reiter (1982a): Z. Phys. *C14*, 47 /Sects.3.2.1,3/
Nachtmann, O., A. Reiter (1982b): Z. Phys. *C16*, 45 /Sects.2.3, 3.2.1/
Nandi, W, W.W. Wada (1980): Phys. Rev. *D21*, 75 /Sect.3.1.3/
Niczyporuk, B. et al. (1981): Z. Phys. *C8*, 1 /Sect.2.2/
Niczyporuk, B. et al. (1982): Z. Phys. *C15*, 299 /Sect.2.1/

Odorico, R. (1980): Z. Phys. *C4*, 113 /Sect.2.4.2/

Particle Data Group (1982): Phys. Lett. *111B*, /Sect.2.4.2/
Pennington, M.R. (1982): Phys. Rev. *D26*, 2048 /Sect.3.2.4/
Pennington, M.R. (1983): Rep. Progr. Phys. *46*, 393 /Preface/
Petermann, A. (1979): Phys. Rep. *53*, 157 /Preface/
Peterson, C. (1980): Z. Phys. *C3*, 271 /Sect.2.2/
Peterson, C., D. Schlatter, I. Schmitt, P.M. Zerwas (1983): Phys. Rev. *D27*, 105 /Sect.2.4.1/
Pi, S.Y., R.L. Jaffe, F.E. Low (1978): Phys. Rev. Lett. *41*, 142 /Sect.3.1.3/
Politzer, H.D. (1973): Phys. Rev. Lett. *30*, 1346 /Sect.1.3/
Politzer, H.D. (1974): Phys. Rep. *14*, 129 /Preface, Sect.2.1/
Politzer, H.D. (1977): Phys. Lett. *70B*, 430 /Sect.3.1.9/

Reya, E. (1981): Phys. Rep. *69*, 159 /Preface/
Richards, D.G., W.J. Stirling, S.D. Ellis (1982): Phys. Lett. *119B*, 193 /Sect.3.2.2/
Ritter, S. (1982): Z. Phys. *C16*, 27 /Sect.2.4.2/
Ross, G.G. (1981): Proceedings of the 21st Scottish University Summer School in Physics, St. Andrews, 1 /Preface/

Sachrajda, C.T. (1982): Southampton University Preprint SHEP 81/82-2 /Preface/
Satz, H. (1975): *Current Induced Reactions*, Lecture Notes in Physics, Vol. 56 (Springer, Berlin, Heidelberg, New York) 49 /Sect.2.4.1/
Saxon, D.H. (1982): Rutherford Appleton Laboratory Report RL-82-096 /Sect.3.1.9/
Schierholz, G. (1979): Proceedings of the SLAC Summer Institute on Particle Physics, Quantum Chromodynamics, 479 /Preface, Sects.3.1.8,9/
Schierholz, G., D.H. Schiller (1979): DESY-Report 79/29 /Sect.3.1.3/
Schierholz, G., J. Willrodt (1980): Z. Phys. *C3*, 125 /Sect.3.1.9/
Schierholz, G. (1981): *Current Topics in Elementary Particle Physics* (Plenum, New York) 77 /Preface/

Schiller, D.H. (1979): Z. Phys. *C3*, 21 /Sect.2.2/
Schiller, D.H., G. Zech (1982): Physica Scripta *26*, 273 /Sect.3.1.2/
Schlatter, D. et al. (1982): Phys. Rev. Lett. *49*, 52 /Sect.3.1.9/
Schneider, H.N., G. Kramer, G. Schierholz (1983): DESY-Report 83-095
    /Sect.3.2.2/
Söding, P. (1981): DESY-Report 81-070, published in Proceedings AIP Particles and
    Fields Division Conference (Santa Cruz) /Sect.3.1.9/
Söding, P., G. Wolf (1981): Ann. Rev. Nucl. Part. Sci. *31*, 231 /Preface/
Soper, D.E. (1983): Z. Phys. *C17*, 367 /Sect.3.1.5/
Sterman, G., S. Weinberg (1977): Phys. Rev. Lett. *39*, 1436 /Sect.3.1.7/
Stevenson, P.M. (1978): Phys. Lett. *78B*, 451 /Sect.3.1.7/
Stevenson, P.M. (1981): Phys. Lett. *100B*, 61; Phys. Rev. *D23*, 2916 /Sect.3.2.4/
Stevenson, P.M. (1982): Nucl. Phys. *B203*, 472 /Sect.3.2.4/
Stueckelberg, E.G., A. Petermann (1953): Helv. Phys. Acta *25*, 499 /Sect.3.2.4/
Symanzik, K. (1970): Commun. Math. Phys. *18*, 227 /Sect.3.2.5/

TASSO Collaboration (1982), (presented by D. Lücke): Proceedings of the 21st In-
    ternational Conference on High Energy Physics, Journal de Physique *43*, C3-67
    /Sect.3.1.9/
Taylor, J.C. (1976): *Gauge Theories of Weak Interactions* (Cambridge University
    Press) /Sect.1.2/
The MARK J Collaboration (1980): Phys. Rep. *63*, 337 /Sect.3.1.9/
t'Hooft, G., M. Veltman (1972): Nucl. Phys. *B44*, 189 /Sect.3.1.1/
t'Hooft, G. (1971): Nucl. Phys. *B33*, 173; *B35*, 167 /Sect.1.1/

Van Hove, L. (1969): Nucl. Phys. *B9*, 331 /Sect.2.4.1/
Vermaseren, J.A.M., K.J.F. Gaemers, S.J. Oldham (1981): Nucl. Phys. *B187*, 301
    /Sect.3.2.2/

Walsh, T.F. (1980): Acta Phys. Austr. Suppl. XXII, 439 /Preface/
Weeks, B.G. (1979): Phys. Lett. *81B*, 377 /Sect.3.1.7/
Wolf, G. (1980): DESY-Report 80/85, Proceedings of the XI. International Symposium
    on Multiparticle Dynamics (Bruges) /Sect.3.1.9/
Wolf, G. (1981): DESY-Report 81-086, Proceedings of the Cargèse Summer Institute
    /Sects.2.1, 3.1.9/
Wolf, G. (1982): Proceedings of the 21st International Conference on High Energy
    Physics, Journal de Physique *43*, C3-525 /Sects.2.1,2, 3.1.9/
Wu, S.L., G. Zobernig (1979): Z. Phys. *C2*, 107 /Sect.2.3/
Wu, S.L. (1981): DESY-Report 81-071, Proceedings of the Topical Conference of the
    1981 SLAC Summer Institute in Particle Physics (SLAC, Stanford, Calif.)
    /Sect.3.1.9/
Wu, S.L. (1981): Z. Phys. *C9*, 329 /Sect.2.3/
Wu, S.L: (1983): DESY-Report 83-007, Proceedings of the Topical Conference of the
    1982 SLAC Summer Institute on Particle Physics (SLAC, Stanford, Calif.)
    /Sect.3.2.2/

Yamamoto, H. (1981): Caltech Preprint CALT-68-836 /Sect.3.1.9/
Yang, C.N., R.L. Mills (1954): Phys. Rev. *96*, 191 /Sect.1.2/

Zee, A. (1973): Phys. Rev. *D8*, 4038 /Sect.2.1/
Zinn-Justin, J. (1975): *Trends in Elementary Particle Theory*, Lecture Notes in
    Physics, Vol. 37 (Springer, Berlin, Heidelberg, New York) 2 /Sect.1.2/
Zweig, G. (1964): CERN Report 8182/TH 401 /Sect.1.1/

W. Hofmann

Jets of Hadrons

1981. 165 figures. VIII, 215 pages
(Springer Tracts in Modern Physics, Volume 90)
ISBN 3-540-10625-1

Today it seems that particle production in almost any type of high energy reactions of elementary particles- electron-positron annihilation, lepton-hadron and hadron-hadron reactions for example – can be attributed to one common mechanism: the fragmentation of partons into jets of hadrons. This book reviews the properties and phenomenology of hadron production in different reactions, using the concept of jets of hadrons as a unifying guide-line.

The properties of jets, as observed in e^+e^- annihilation, in lepton-nucleon reactions and in decays of the Υ meson discovered recently are summarized and are compared to predictions of the quark parton model. Special emphasis is placed on the discussion of jets occurring in hadron-hadron reactions where partons are scattered at large angles and give rise to the production of particles with unusually high transverse mementa.

The discussion is kept at an elementary and intuitive level throughout the book. The authors aim not so much to provide an exhaustive compilation of data to present the basic concepts involved in jet- or parton physics.

Springer-Verlag
Berlin
Heidelberg
New York
Tokyo

Zeitschrift für Physik C

Particles and Fields

Subscription Information is available from your bookseller or directly from Springer-Verlag, Wissenschaftliche Information Zeitschriften, Postfach 105 280, D-6900 Heidelberg, FRG

Orders from North America should be addressed to: Springer-Verlag New York Inc., Journal Sales Dept., 44 Hartz Way, Secaucus, NJ 07094, USA

 Europhysics Journal

ISSN 0170-9739

Title No. 288

Editors in Chief: **G. Kramer,** Hamburg; **H. Satz,** Bielefeld

Editors: **R. Barbieri,** Pisa; **T. Ferbel,** Rochester; **K. Fujikawa,** Hiroshima; **P. Hasenfratz,** Genf; **K. Kajantie,** Helsinki; **A. Krzywicki,** Orsay; **P. Söding,** Hamburg; **B. Stech,** Heidelberg; **J. C. Taylor,** Cambridge; **F. Wilczek,** Santa Barbara

Aims and Scope:
Zeitschrift für Physik C, **Particles and Fields** is devoted to the experimental and theoretical investigation of elementary particles. In view of the steadily growing interplay of theory and experiment in this field, particular emphasis is given to a clear and complete presentation of research.

The topics covered include:
- Experimental and theoretical particle physics
- Structure of elementary particles
- High energy processes
- Strong, electromagnetic and weak interactions
- Symmetry principles
- Unification schemes
- S-matrix theory
- Quantum field theory
- Lattice field theory

Special features: Rapid publication, no page charge. Language of publications is English.

Zeitschrift für Physik appears in three parts: A: Atoms and Nuclei; B: Condensed Matter; C: Particles and Fields Each part may be ordered separately.
Coordinating editor for Zeitschrift für Physik, Parts A, B and C: O. Haxel, Heidelberg

Springer-Verlag
Berlin
Heidelberg
New York
Tokyo